Progress in Molecular and Subcellular Biology 5

Progress in Molecular and Subcellular Biology

5

With Contributions by
T. Arpa-Gabarro · W. Flamenbaum · N. Gabel
P. Gund · R. J. Hamburger · J. S. Kaufman
C. Reiss · I. Schuster · J. Schwartz

**Edited by: F. E. Hahn, H. Kersten,
W. Kersten, W. Szybalski**

**Advisors: T. T. Puck, G. F. Springer,
K. Wallenfels**

Managing Editor: F. E. Hahn

Springer-Verlag   Berlin   Heidelberg   New York   1977

Professor Fred E. Hahn, Ph.D.
Department of Molecular Biology
Walter Reed Army Institute of Research
Washington, D.C. 20012

With 56 Figures

ISBN 3-540-08192-5 Springer-Verlag Berlin Heidelberg New York
ISBN 0-387-08192-5 Springer-Verlag New York Heidelberg Berlin

The Double Helix Revisited: Watson and Olby

Fred E. Hahn

In 1968, James D. WATSON startled the scientific community, as well as the laity, with his publication of *The Double Helix, A Personal Account of the Discovery of the Structure of DNA*. There had not before been a debunking of science just like that since Sebastian BRANT wrote (BASEL, 1494) in *Narrenschiff*:

> *Damit ich nit vergess allhie*
> *den grossen B'schiss der Alchemie.*

WATSON's book received numerous reviews in 1968 (BRONOWSKI; CHARGAFF; Editorial; LEAR; LWOFF; MEDAWAR; MERTON; MORRISON); STENT (1968) even wrote a review of such reviews.

Six years later, a detailed and scholarly account of the pre-history and history of the development of the double helical DNA structure was published: OLBY's book (1974), *The Path to the Double Helix*. As an added attraction, it contains a brief Foreword by Francis H.C. CRICK in which he addresses himself (to my knowledge for the first time in published writing) inter alia to WATSON's book.

WATSON's tale begins in 1950 with his youthful peregrinations, which eventually landed him at the Cavendish to team up with CRICK; it leaves it to the scientifically educated reader to supply more than a decade of science history, leading up to the point at which the determination of the structure of DNA could be perceived as a key task in the life sciences. Only en passant does WATSON mention AVERY's discovery of DNA being the bacterial type-transforming principle. CHARGAFF's rules of regularities in DNA's base composition are presented in one paragraph.

Evidently, WATSON was not chiefly concerned with the history of science, although he tells us about the impact of his learning of the Hershey-Chase experiment in which he saw "a powerful new proof that DNA is the primary genetic material," and we also read that CHARGAFF's rules at first "did not ring a bell" [sic] with CRICK who soon, however, came around to considering them "a real key" to the DNA structure. The forte of WATSON's story is personalities and their interrelationships. His book is, indeed, a "personal account." Moreover, personal peeps in the story go beyond what is necessary to understand the events culminating in the discovery of the DNA structure.

Enough has been written in reviews on the propriety, or absence
thereof, in WATSON's dealing with those who were involved in the
story of the double helix. His is a rather unmerciful sense of humor.
After they both had become the famous Dioscuri, Drs. WATSON and CRICK
were in the habit of projecting at scientific meetings (!) candid
photographs, which they had surreptiously taken of each other, and
which showed the victims in less than flattering poses. And WATSON's
dealing with people in his story is invariably candid and less than
flattering. Whether one finds this shocking or acceptable depends on
one's manners and literary tastes.

Because the actual story happened 25 years ago, WATSON's gossip is
somewhat dated in this year 1976. The ultimate significance of *The
Double Helix* lies neither in its contribution to the history of science nor in its somewhat unpleasant sense of humor when dealing with
people who, if still alive, are by now in their comfortable middle
ages. But from time to time, there have appeared works of fiction
such as Upton SINCLAIR's *The Jungle* or Bud SCHULBERG's *What Makes
Sammy Run*? that suddenly bring into spotlit focus the inner workings
of one or the other sub-society. In this sense, Sinclair LEWIS'
Arrowsmith was an early forerunner of *The Double Helix*.

Two features seem to set *The Double Helix* apart from this genre of
literature. Firstly, it has a factual basis and deals with actual
people and events (although the news of the priority of d'HERELLE's
discovery of bacteriophage in *Arrowsmith* is authentic enough).
WATSON originally planned to write his story in the form of a roman à
clef. Secondly, it is a success story. While the literary heroes of
works of social critique ultimately resign (*Arrowsmith*; IBSEN's *An
Enemy of The People*; CHEKOV's *The Three Sisters*) or perish (MILLER's
Death of a Salesman), WATSON and CRICK succeeded probably beyond
their fondest expectations at the outset of their quest. Readers
might be tempted to give the book a different subtitle: *How to Succeed in Contemporary Science*.

Social critique in WATSON's book is inadvertent. There is nothing
overt or intentional about it. And why should this be? Drs. WATSON
and CRICK won over their competitors and became celebrities and Nobel
Laureates who only rarely are still lambasted by inveterate adver-
saries of molecular biology (CHARGAFF, 1974). According to *The
Double Helix*, WATSON and CRICK, when setting out to discover the
structure of DNA, were propelled by their competitive drive for priority, especially over Linus PAULING who looms throughout the book as
the Great Menace although (like Juarez in WERFEL's *Juarez and
Maximilian*) he never appears on stage. The race for the building of
the DNA model has its parallels in the race for the building of the
atomic bomb or of the moonship. This is the way things are nowadays.
Those who believe naively that scientists are unselfishly dedicated
to investigating the phenomena and laws of nature (see Eve CURIE's
biography of Maria Curie) find themselves confronted in WATSON's tale
with the stark secularization of the scientific establishment and
with its accomplished shift in emphasis from the advancement of science to the advancement of scientists.

As the chronicle unfolds, we read about WATSON's hope that his sister might induce WILKINS to associate WATSON closely with the X-ray work on DNA, about the use of PAULING's son as an agent of information on the progress in his father's work and about the most timely receipt of a privileged communication of WILKINS and FRANKLIN through an administrative indiscretion, albeit not of WATSON's and CRICK's making. The book does not indicate that its author saw anything wrong with such catch-as-catch-can wrestling for success. If an end-justifying-the-means ethic is rampant in the contemporary scientific world--and how can one overlook that it is?--WATSON should not be criticized for having told his inside story of the discovery of the DNA structure but might have deserved a Pulitzer Prize for reporting it in all his amoral and naive candor.

This then is the point: social critique is not offered or implied, but it is evoked in the mind of the reader who is led to ask, Is science really like that? For this, one can forgive the rendition of the story its lack of taste and its cattiness; for all that it makes for easy and amusing reading.

Finally, the reader comes to understand that there was a genuine element of creativity in assembling the DNA structure, if one regards creativity in science as the ability to reason correctly on the basis of incomplete information. WATSON's flash of insight into the structural meaning of CHARGAFF's rules and into the genetic significance of complementarity of the two DNA strands was a moment of creative intuition comparable to KEKULÉ's vision of the benzene ring. Of such rare moments is made the element of happiness in the life of a scientist most of which is spent in respectable drudgery, as in SHAKESPEARE's lines, "Bubble, bubble, toil and trouble."

The Path to the Double Helix begins in the 1920s by tracing the genesis of the concept of macromolecules and the development of hydrodynamic and optical methods for their study. OLBY then proceeds to recount the development of X-ray fiber analysis, which, for DNA, was originally worked out by the Leeds group under ASTBURY who, with BELL, published in 1938 the remarkable DNA model, shown in Fig. 1, in which the bases are stacked at 3.3 Å intervals.

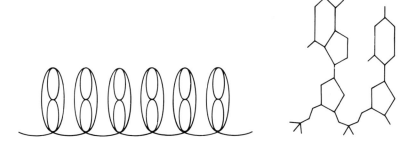

Fig. 1. Single-helical DNA model, with the bases stacked at 3.3 Å intervals, proposed in 1938 by ASTBURY and BELL

We cannot reiterate OLBY's painstaking recounting of the gradual evolution of molecular genetics, which was stymied by the general belief
in the protein nature of genes and by the sway held by the tetranucleotide hypothesis of the DNA structure. Noteworthy is OLBY's
treatment of the "chemistry of virus genes." I have always thought
that the isolation of tobacco mosaic virus by STANLEY (1935) was in
fact, although not in name, the moment of birth of molecular biology.
Anyone who was active in science during the late 1930s remembers the
excitement and controversy elicited by the idea of an autoreplicative
virus molecule. About a decade later, the author of this review presented and published a Heidelberg Lecture (HAHN, 1948) in which he
addressed himself to the seeming dualism of crystallized plantvi-
ruses: molecules on the one hand and replicative biologic entities on
the other; he also suggested a "matrix hypothesis" according to which
the *nucleic acid moiety was the genetic determinant of the identical
replication of plant viruses.*

Proof that DNA is the physical carrier of bacterial heredity had come
from the type-transformation study of AVERY, MACLEOD and MCCARTY
(1944). Although OLBY gives broad coverage to this discovery and its
pre- and subsequent histories, he, like other scholars, cannot explain the unbelievable lethargy of the scientific community presented
with the AVERY result. Certainly, there existed a strong pro-protein
bias in the emerging field of biochemical genetics as well as the
stranglehold of the tetranucleotide hypothesis on thoughts concerning
a genetic role of DNA. But all this cannot quite account for the
lack of recognition of the fundamental significance of AVERY's finding.

Much later (1967), Arne TISELIUS, then head of the Nobel Chemistry
Committee, would say: "That AVERY never received the prize is lament-
able and had he not died when he did, I think he would almost certainly have gotten it. However, he was an old man when he made his
discovery. Most people believed he was right, but there was still
some doubt. By the time his achievement was confirmed, he had died.
This is one case I regret very much (LITELL, 1967). It is true that
AVERY was 67 years old when he published his masterpiece, but he
lived another 11 years, i.e., three years beyond the publication of
the HERSHEY-CHASE experiment (1952), which demonstrated that phage
DNA was the physical carrier of bacterial virus infectivity. STENT's
(1972) suggestion that AVERY's discovery was not appreciated by the
biologic community because it was "premature" may come close to the
truth; at that, it is a sad commentary on the history of scientific
ideas.

OLBY, finally gives a detailed account of the discovery of constant
base ratios in DNA by CHARGAFF. Not only did these analytical results remove the conceptual obstacle of the tetranucleotide hypothesis, i.e., of the proposition that DNA was a monotonous and, hence,
informationally trivial succession of identically repeated tetranucleotide subunits, but they established *laws of constant proportions*
for the base composition of DNA (CHARGAFF et al., 1951). It is difficult to understand that the demonstration of these stoichiometric

relationships did not elicit the conditioned response of analytical chemists, viz., to infer that compositional regularities mean that subunits present in unitary and constant molar ratios are bound to each other in the analyzed molecule. Even the chemist Kurt G. STERN, who came close to the truth by entertaining the idea of hydrogen bonding between bases in DNA, did not recognize the connection between constant base ratios and the scheme of base pairing (so OLBY reports).

From 1951, WATSON's and OLBY's accounts run parallel with the exception of an Epilogue to WATSON's book with an esprit de l'escalier vindication of Rosalind FRANKLIN who had died in 1958 or the Conclusion of OLBY's book with commentary on, inter alia, the DNA model and the Central Dogma.

OLBY has written the definitive scholarly account of the development and discovery of the DNA structure, the currents of conceptual and informational influences, the manifold personal interactions, and the temporary blocks and blind alleys in the tortuous path to the double helix. He includes numerous quotations from personal letters (many received by himself), which are not generally available for reference, in addition to quoting the pertinent scientific literature. This is a handsome example of writing history of science and also contains enough biographical data and capsule characterizations of the major dramatis personae, so that the people who acted and interacted in the quest for the DNA structure come alive.

Finally, in his Conclusion, OLBY gives one page to a topic that he calls "The Uniqueness of Discovery." He presents a credible argument that several groups of workers might sooner or later have arrived at the correct structure of DNA by continuing their progressive scholarly investigations. However, if the DNA structure had been determined as the result of such efforts, the story of its discovery would be less dramatic. In other key areas of molecular biology, e.g., the mechanism of protein biosynthesis, the very large number of individual scientific contributions scattered in the literature presents a picture of collective discovery in which only few contributions stand out as truly distinctive advances. The fact that the DNA structure emerged from the efforts of WATSON and CRICK in one single and unique burst of scientific creativity, comparable to an act of artistic creation, makes the discovery noteworthy. For this, WATSON and CRICK with WILKINS received the Nobel Prize in 1962. Sten FRIBERG, chairman of the Nobel Medicine Committee, once said: "The science prizes are never awarded for a lifetime's work; they are awarded for one achievement, one discovery" (LITELL, 1967). WATSON's and OLBY's books describe such a discovery, one from the viewpoint of intense personal involvement and the other from the detached vantage point of the historian. Both books are worthy of being read, preferentially side-by-side.

References

ASTBURY, W.T., BELL, F.O.: Some recent developments in the X-ray study of proteins and related structures. Cold Spring Harbor Symp. Quant. Biol. 6, 109 (1938).

AVERY, O.T., MacLEOD, C.M., McCARTY, M.: Studies on the chemical nature of the substance inducing transformation of pneumococcal types. J. Exp. Med. 79, 137 (1944).

BRONOWSKI, J.: Honest Jim and the tinker toy model. Nation 206, 381 (1968).

CHARGAFF, E.: A quick climb up Mount Olympus. Science 159, 1448 (1968).

CHARGAFF, E.: Building the tower of Babble. Nature (Lond.) 248, 776 (1974).

CHARGAFF, E., LIPSHITZ, R., GREEN, C., HODES, M.E.: The composition of the desoxyribonucleic acid of salmon sperm. J. Biol. Chem. 192, 223 (1951).

Editorial: Central Dogma, right or wrong? Nature (Lond.) 218, 317 (1968).

HAHN, F.: Probleme der Virusforschung und ihre naturwissenschaftlichen Aspekte. Heidelberger Vortrage 10, 5. Heidelberg: F.H. Kerle 1948.

HERSHEY, A.D., CHASE, M.: Independent functions of viral protein and nucleic acid in growth of bacteriophage. J. Gen. Physiol. 36, 39 (1952).

LEAR, J.: Heredity transactions. Saturday Rev. 51, 36 (1968).

LITELL, R.J.: The Nobel establishment: A rare glimpse. Sci. Res. 2 (10), 48 (1967).

LWOFF, A.: Truth, truth, what is truth (about how the structure of DNA was discovered)? Sci. Am. 219, 133 (1968).

MEDAWAR, P.B.: Lucky Jim. New York Review of Books 10 (6), 3 (1968).

MERTON, R.K.: Making it scientifically. The New York Times Book Review (February 25), 1 (1968).

MORRISON, D.: Human factor in a science first. Life Magazine 64 (9), 8 (1968).

OLBY, R.: The Path to the Double Helix. Seattle: Univ. Wash. Press 1974.

STANLEY, W.M.: Isolation of a crystalline protein possessing the properties of tobacco-mosaic virus. Science 81, 644 (1935).

STENT, G.S.: What are they saying about Honest Jim? Q. Rev. Biol. 43, 179 (1968).

STENT, G.S.: Prematurity and uniqueness in scientific discovery. Sci. Am. 227 (6), 84 (1972).

WATSON, J.D.: The Double Helix. A Personal Account of the Discovery of the Structure of DNA. New York: Atheneum 1968.

Contents

List of Contributors

T. ARPA-GABARRO, Foundation Curie, Institut du Radium, Section de
   Biologie, 15 rue Clemanceau, F-Orsay 91405, France

WALTER FLAMENBAUM, Chief, Renal Section, Boston Veterans Administration
   Hospital, 150 South Huntington Avenue, Boston, Massachusetts 02130,
   USA

NORMAN W. GABEL, Planetary Biology Division, National Aeronautics and
   Space Administration, Ames Research Center, Moffett Field,
   California 94035, USA

PETER GUND, Merck, Sharp and Dohme Research Laboratories, Building 80,
   P.O. Box 2000, Rahway, New Jersey 07065, USA

FRED E. HAHN, Department of Molecular Biology, Walter Reed Army
   Institute of Research, Washington, D.C. 20012, USA

ROBERT J. HAMBURGER, Renal Section, Boston Veterans Administration
   Hospital, 150 South Huntington Avenue, Boston, Massachusetts 02130,
   USA

JAMES S. KAUFMAN, Department of Nephrology, Walter Reed Army Institute
   of Research, Washington, D.C. 20012, USA

CLAUDE REISS, Foundation Curie, Institut du Radium, Section de
   Biologie, 15 rue Clemanceau, F-Orsay 91405, France

INGEBORG SCHUSTER, Sandoz Forschungsinstitut, Abt. für Biochemie,
   A-1235 Wien, Austria

JOHN SCHWARTZ, Department of Nephrology, Walter Reed Army Institute of
   Research, Washington, D.C. 20012, USA

Thermal Transition Spectroscopy: A New Tool for Submolecular Investigation of Biologic Macromolecules

Claude Reiss and T. Arpa-Gabarro

I. Introduction

The investigation of local properties of biologic macromolecules is
both very appealing and quite difficult. The knowledge of local composition and conformation is required, if fundamental processes of
molecular biology are to be understood. For instance, storage, processing, or transfer of genetic information at the gene level, or the
elementary processes occurring in enzymatic reactions depend on submolecular characteristics of nucleic acids or proteins. That investigations of local properties is difficult results from the fact that
biologic macromolecules cumulate their polymeric character with a
rather impressive chemical complexity.

Methods recently became available for local studies of biologic particles. For instance, sequences of small stretches of DNA, RNA, or
small proteins have been established. X-ray, neutron scattering of
crystallized proteins, or tRNAs have yielded precise information on
their submolecular structure, pertinent at least in the solid, organized state. To what extent these structures remain in isolated particles can, however, not be stated, since the diffusion studies as
performed nowadays yield only overall conformation parameters. Magnetic resonance, absorption, and fluorescence studies of solutions
can, in some instances, give precise information on submolecular
structures; these methods are, however, restricted to particular
cases, or have to rely on the presence of artificially introduced labels, which are likely to perturb the local environment. Electron
microscopy allows the study of local properties of isolated molecules
of DNA, but the sample-handling prior to the observation may give
rise to artifacts that are difficult to control.

Thermal unfolding of biologic macromolecules may, in principle, be a
way to study their submolecular characteristics, even in solution.
For instance, it is well known that nucleic acids and proteins denature when heated; strand separation occurs in DNA, and helical parts
in proteins or tRNA unfold.

The quantitative measurement of thermal denaturation has been used
for overall estimates, such as mean base composition of DNA or average helical content of proteins. In principle at least, the possibility of using thermal unfolding for submolecular investigation

should exist. For DNA, the thermal stability of AT pairs is much
lower than that of GC pairs. Therefore, the AT or GC clusters in a
DNA molecule, which can be seen by electron microscopy of virus DNA,
for instance (INMAN, 1966), should denature at different temperatures.
Similarly, a protein may contain several α-helical segments, each
denaturing in a distinct step. However, on examining the literature
dealing with thermal studies of biologic macromolecules, no multistep
patterns can be seen in the heat denaturation recordings; instead,
smooth and rather broad transitions from the native to the denatured
state are the rule.

A very few exceptions do, however, exist (FALKOW and COWIE, 1968;
BERNARDI et al., 1970). These, and the theoretical studies published
by several authors (CROTHERS and KALLENBACH, 1966; FIXMAN and ZEROKA,
1968; REISS et al., 1966), prompted us to reexamine this situation.
Surprisingly, throughout the literature, thermal denaturation of biologic material is conducted in very similar ways. The temperature
program to which the samples are subjected is usually a linear one,
with slopes ranging from 0.5°C/min up to 2°C/min. Obviously, such
experimental conditions are incapable of showing unambiguously fine
structural features, if any, in the melting curve. However carefully
the measurements of the denaturation monitoring parameters are performed, such experiments do not take note of the thermodynamic state
of the sample under investigation. Neither thermal nor phase equilibrium is reached within the sample during the measurements, so that
the results are not very accurate, poorly reproducible, and inadequate for comparison with results from theoretical investigations,
which assume well-defined thermodynamic conditions.

A. Heat Denaturation at Equilibrium

We found that the equilibrium requirements just outlined are best met
when the sample is subjected to a *stepwise incrementation of the temperature*. In each step, the temperature of the sample is kept constant for a time sufficient not only for thermal equilibrium, but
also for phase equilibrium to be reached. Such a procedure offers
several advantages over the customary, linear time/temperature profile:

1. Measurements are performed at full equilibrium of the sample; the
   recorded data are thermodynamically meaningful and reproducible.
2. Since the sample is at equilibrium, the measurement may be repeated n times, which is known to increase the accuracy (the sig-
   nal/noise ratio is improved by $\sqrt{n}$).
3. Using an analytical instrument of given performance, a minimum
   temperature step ΔT can be defined that states the minimum temperature increment yielding a meaningful difference of the denaturation monitoring parameter. The improved accuracy brought
   about by the iteration procedure allows a reduction of ΔT, so that
   the denaturation curve can be described with more points per unit
   temperature. Possible fine structural features in the denaturation curve may thus show up.
4. High accuracy temperature and monitoring parameter measurements

allow a meaningful computation of the derivative of the latter
with respect to the former. Any fine structural feature in the
denaturation behavior of the sample is strongly emphasized in the
derivative representation, since the inflection points in the
usual sigmoidal denaturation curve show up as extrema in the derivative curve.

B. Experimental Set-up

It is not our purpose to give a detailed description of an apparatus
that meets the specifications just outlined. We will only draw its
general features and indicate some recent improvements over the set-
up described earlier (REISS and MICHEL, 1974).

As an example, we assume that the denaturation monitoring parameter
is the absorbance variation at a given wavelength (hypochromicity).
Any other parameter sensitive to the denaturation state of the molecule to be studied (NMR, fluorescence, viscosity, etc...) can be
handled in a similar way. The large amount of data that are to be
gathered if the preceding procedure is adopted, i.e., small temperature increments, numerous repetitions of the monitoring parameter at
a given temperature, can best be handled by a fully automated system,
including a spectrophotometer, a thermostatic bath, a microprocessor,
and a data storage device (punched tape, minicassette, disc, etc...).

C. Spectrophotometer

A double beam spectrometer is advisable, owing to its improved stability over single beam devices. Special attention should be payed
to the photometric accuracy, which may be as good as 10^{-4} A for an
initial absorbance of ~ 0.5 A. Accessories should include repetitive
scan and automatic cuvette change, fitted with heated cuvette holders.
Digital output is provided by the most modern instruments. The stur-
diness of the instrument is not the last quality required, because
long-lasting experiments with many scans and cuvette changes require
optical, electronic, and mechanical parts that will remain trouble-
free under heavy-duty conditions.

D. Temperature Control

We found that the most accurate thermostation of the sample is provided by circulating a thermostated liquid (usually water) around the
spectrophotometer cuvettes. Thermostated water is flushed through
jacketed quartz cuvettes and through the cuvette holder; the temperature inside the sample chamber, housed in the quartz cuvette, is thus
held at a stable value with respect to that of the thermostated water
bath. No temperature gradient in a 0.5 ml sample chamber could be
detected, using an interferometric set-up that would have revealed
gradients of less than 10^{-3} C.

Water thermostation is provided by an ultrathermostat, with a pump
and a cold source (tap water, cryostat, etc...). The temperature-

sensing device housed in the thermostat is a wire-wound resistor (platinum, 100Ω at 0°C). Temperature control can be provided by an analog, null-balance proportional controller, or better via a digital voltmeter with direct temperature read-out. In the latter case, the difference between the actual temperature and the set value is con- verted by a microprocessor (see below), into a pulse train; each pulse fires a triac circuit, feeding the electric heater of the bath. This procedure will not cause any temperature "overshoot," although the temperature comes very rapidly to the set value. With the equipment we have built recently in our laboratory, the temperature in the ul- trathermostat is constant over hours to $\pm 1.5 \times 10^{-3}$ C, whereas the temperature fluctuations in the sample housing are beyond the sensi- tivity of the voltmeter, 10^{-3}°C.

E. Time-temperature Program and Interfacing the Spectrophotometer with the Data Storage Unit

These two functions, together with the digital temperature control just mentioned, are best performed via a single microprocessor unit.

The chronology of events is as follows. At the beginning of the ex- periment, the temperature is held constant for a fixed time. Then a measurement cycle is performed. Each cycle consists of the measure- ments of the temperature, followed by the absorbance recording (aver- aged on n values) for every cuvette and every preset wavelength; the cycle ends with another measurement of the temperature. Then the temperature consignment is incremented by a set value and a time interval is allowed before the next cycle starts (this procedure goes on until the consignment temperature reaches a final set value).

The microprocessor clock delivers the start signal to the spectropho- tometer, which then performs automatically the cuvette and wavelength changes, and measures the absorbances. The temperature and absorbance measurements, together with the proper addressing signals, are fed through the microprocessor to a data storage system--punched tape, minicassette, etc...

Waiting times, temperature increment, initial and final temperatures are memorized in the microprocessor prior to the start of the experi- ment.

F. Data Treatment

An actual experiment with six samples and three wavelengths, with a temperature step of 0.05°C over 15°C, lasts 15h and results in the recording of some 10^6 bits of information; the absorbances are to be filed in as many functions as there are cuvette-wavelength couples, so that each function corresponds to the variation of absorbance of a given sample at a given wavelength. Each function has to be processed, that is "debugged," smoothed, derived with respect to temperature, and plotted. When necessary, deconvolution of the melting profile into a set of elementary curves-gaussian for instance-can be performed

by a curve-fitting procedure. A rather powerful computer is required
for the curve-fitting procedure. It is highly advisable to process
the data on line, via a minicomputer.

II. Results

The samples corresponding to the melting curves (Figs. 1-10) shown
below were gifts from several laboratories:

λ DNA: Dr. P. Goldenstein (Laboratoire de Microbiologie, Gif); $\emptyset80$
DNA: Dr. M. Crepin (Institut Pasteur, Paris); PM2 DNA: Dr. M. Toutain
(Institut de Recherche sur le Cancer, Villejuif); T5 DNA: Dr. Legault-
Demare (Institut du Radium, Orsay); T7 DNA: Dr. G. Butler-Brown
(Portsmouth); Physadrum DNA: Dr. J. Baldwin; "lac" fragment from λ
plac DNA: Dr. A. Spasky, (Institut Pasteur, Paris); left 40% of λ ge-
nome and pooled fragments of λ plac, cut by Eco R_1: Drs. P. Tiollais
and A. Fritsch (Institut Pasteur, Paris); Adenovirus II: Dr. P.
Boulanger (Lille University).

Prior to melting, the samples were subjected to extensive dialysis in
dilute SSC buffer (SSC = 0.15 M NaCl; 0.015 M Na citrate), as indicated
in the figures. Dialysis proceeded over 36 h, with three changes of
the buffer. The amount of DNA required for a melting curve is 7-10 mg.

Processing of the data (derivation, smoothing...) is performed as described elsewhere (REISS and MICHEL, 1974).

The experimental set-up used in most experiments includes a DMR10,
double-beam spectrophotometer (Carl ZEISS, Oberkochen), with an automatic cuvette changer ZV10. The temperature programming and interfacing electronics were built by DATEC, Paris. Processing of the data
was performed as described (REISS and MICHEL, 1974), on a UNIVAC 1110
computer (Centre de Calcul, Universite de Paris-Sud, Orsay).

A. The Melting Modes

Electron microscopic studies (INMAN, 1966) demonstrate clearly that
functional DNAs are heterogeneous in base composition. The purpose of
building the high resolution device for thermodenaturation was to investigate whether such compositional heterogeneities show up in the
denaturation profile. The results just described demonstrate that
features do indeed appear in the melting curves, so the question that
immediately arises concerns their origin: are the melting modes finger-
prints of intramolecular events upon melting, and if so, do they reflect the known internal heterogeneity?

B. Melting Modes Reflect Intramolecular, Well-localized Events

If population heterogeneities (i.e., base-pair sequences differing
among the DNA molecules present) or conformation particularities among

6

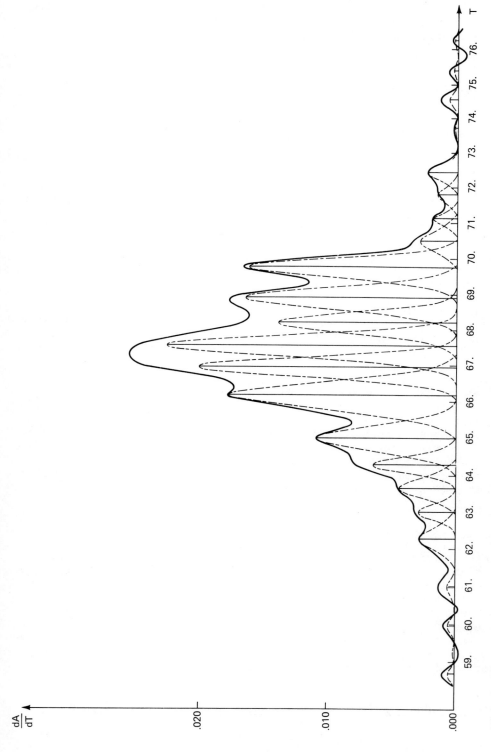

Fig. 1. PM$_2$ DNA, nicked, in SSC x 10-1, temperature step 0.1°C, equilibrium delay 2 min

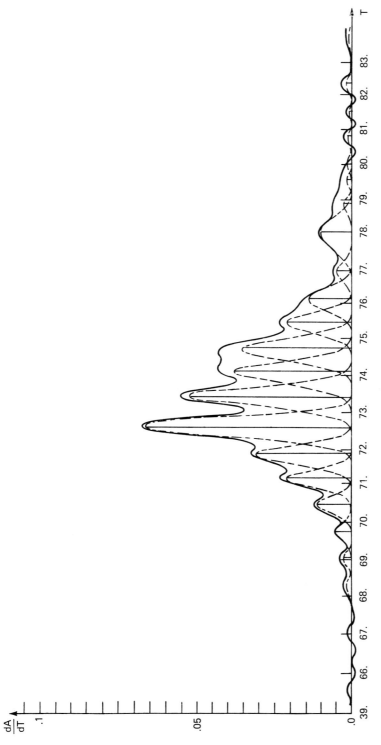

Fig. 2. T$_5$ DNA, in 10−3 M NaCl, 10−3 M Na citrate, temperature step 0.1°C, equilibrium delay 2 min

8

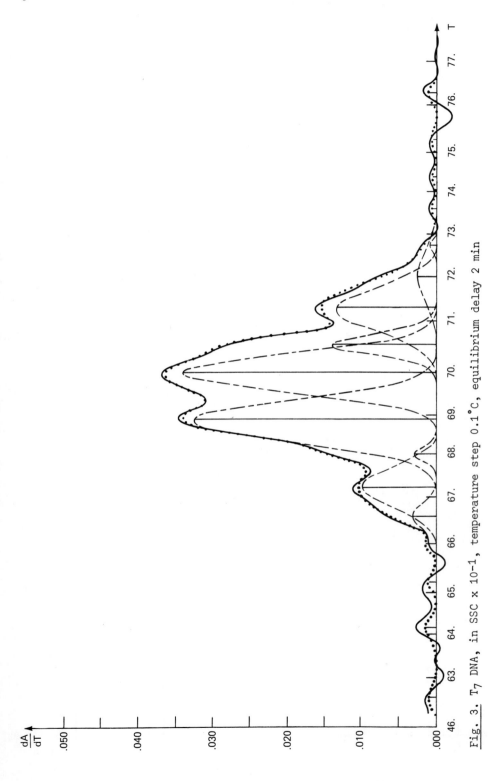

Fig. 3. T$_7$ DNA, in SSC × 10^{-1}, temperature step 0.1°C, equilibrium delay 2 min

9

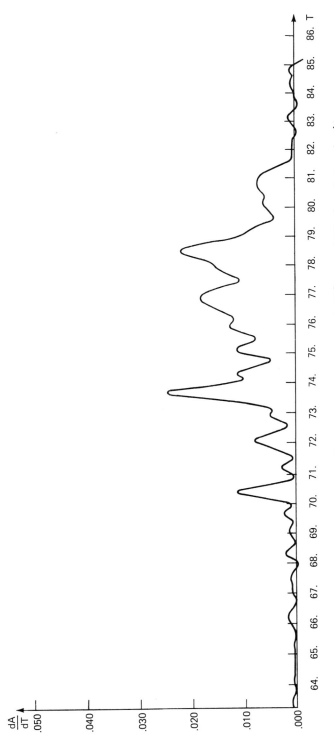

<u>Fig. 4.</u> Physarum Polycephalum DNA, in SSC/10, temperature step 0.05°C, equilibrium delay 1 min

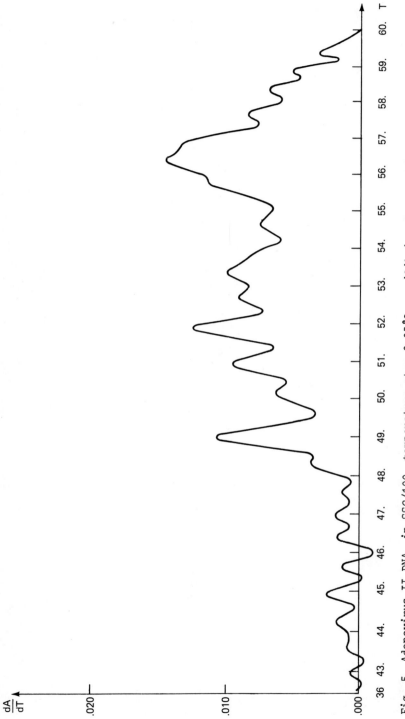

Fig. 5. Adenovirus II DNA, in SSC/100, temperature step 0.05°C, equilibrium delay 1 min

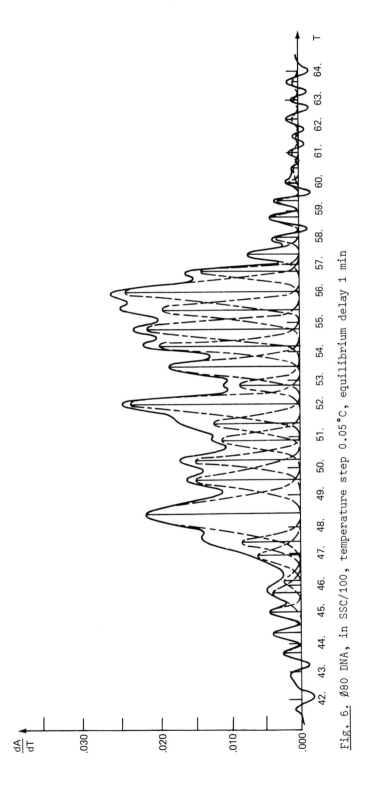

Fig. 6. Ø80 DNA, in SSC/100, temperature step 0.05°C, equilibrium delay 1 min

12

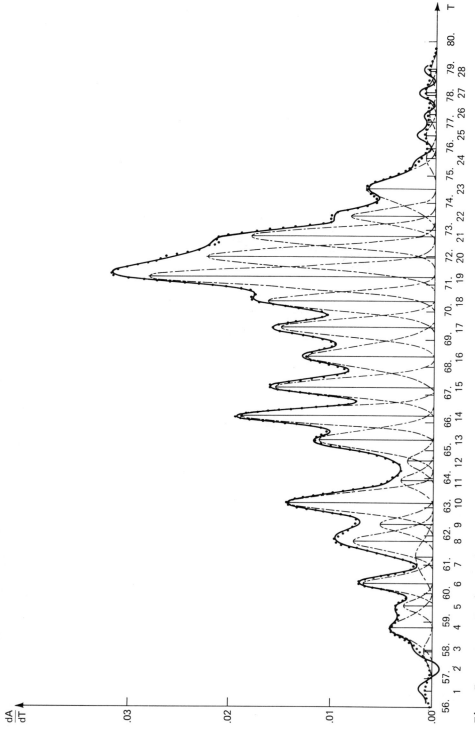

Fig. 7. λ plac DNA in SSC/20, temperature step 0.05°C, equilibrium delay 1 min

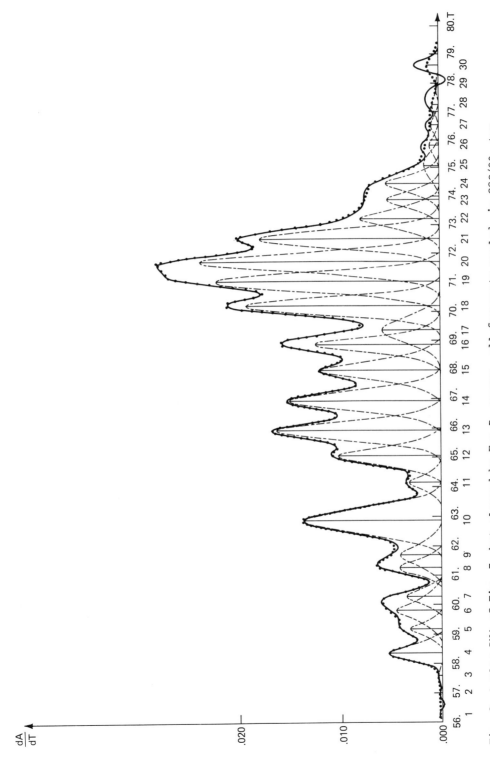

Fig. 8. λ plac DNA of Fig. 7, but cleaved by Eco R₁ enzyme, all fragments pooled, in SSC/20, temperature step 0.05°C, equilibrium delay 1 min

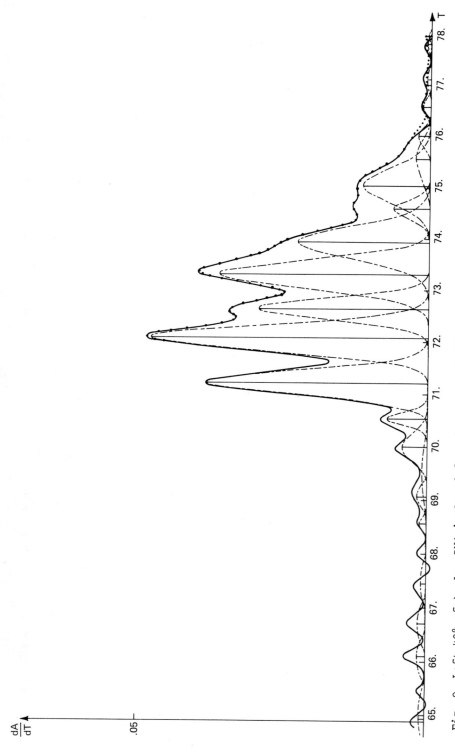

Fig. 9. Left 40% of λ plac DNA isolated from fragments of DNA corresponding to Fig. 8, in SSC/20, temperature step 0.05°C, equilibrium delay 1 min

15

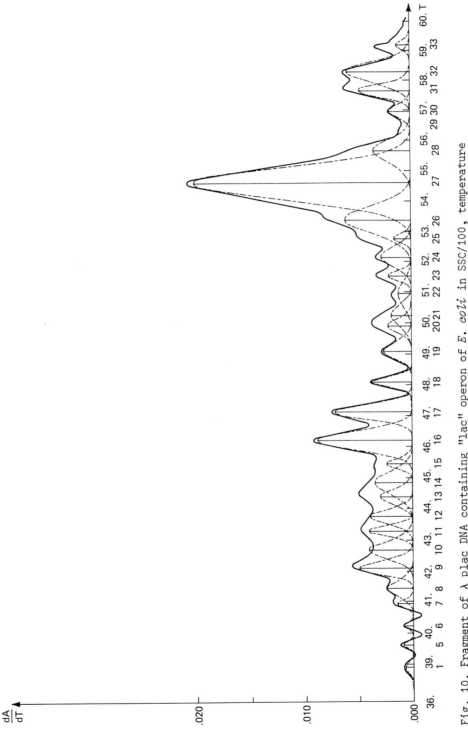

Fig. 10. Fragment of λ plac DNA containing "lac" operon of *E. coli* in SSC/100, temperature step 0.05°C, equilibrium delay 1 min

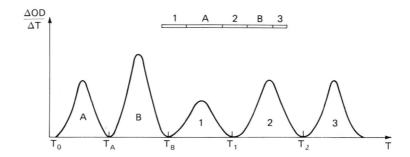

<u>Fig. 11.</u> Assumed melting profile of DNA with thermosome topology shown in caption

the molecules (i.e., secondary structures differing from one molecule to the other, or circularization of part of the population, etc...) were present, each species could give rise to a particular melting profile. For phage DNA, intermolecular heterogeneity can be ruled out. The bacteriophages used were genetically well characterized; buoyant density studies of the DNA showed a single-banding species; factors known to affect the conformation of DNA (polyvalent ions and proteins in particular) were carefully eliminated by appropriate treatments. Electron microscopic studies on λ DNA (INMAN, 1966) also argue against intermolecular heterogeneity. All DNA molecules present in a given solution are of similar conformation and composition.

We are thus left with the fact that the modes originate in intramolecular heterogeneities. Two kinds can be foreseen: localized heterogeneities and those spread all over the molecule. As to the latter, one could, for instance, imagine processes that, as denaturation progresses, induce transconformations of the entire DNA molecule, and thus interrupt the melting process until a higher temperature has been reached. This is obviously not the case here, since the cut-off DNA of phage λ melts almost identically to the native species.

We are left with the conclusion that the melting modes in DNA reflect intramolecular, well-localized phenomena. The exact nature of this phenomenon will be discussed in the next section.

C. The Physical Origin of Melting Modes

Thermodynamics of DNA melting have been described in detail in a set of rather sophisticated papers (CROTHERS and KALLEBACH, 1966; FIXMAN and ZEROKA, 1968; REISS et al., 1966). These provide evidence that internal composition heterogeneity can, indeed, explain the occurrence of melting in several steps. For instance, if it is assumed that the DNA is a model block copolymer, where AT- and GC-rich blocks of given length alternate in a regular fashion, mono- or bimodal melting is predicted, according to the block length (POLAND and SHERAGA, 1969). However, for natural DNA, even the most advanced theories (LEHMAN, 1967) predict only broad, featureless melting curves. Rather than

trying to adapt these theories to explain our results, we will develop a qualitative argument showing how multistep melting can occur.

The thermodynamic behavior of a nucleic acid molecule may be discussed
with the aid of an associated melting-temperature (m.t.) profile in
which a series of base pairs is represented by a corresponding series
of points on a graph that correlates the *actual* m.t. (ordinate) of
each base pair with its distance from the end of the molecule (abscissa). As outlined above, we make the important assumption that
melting occurs under equilibrium conditions. The actual m.t. of a
base pair is then the temperature at which its denaturation occurs *in
situ*. Extramolecular parameters (other than temperature) are not
likely to be directly involved in the appearance of polyphasic melting, as demonstrated by the fact that the results reported above were
obtained under constant characteristics of the aqueous medium surrounding the molecules: chemicals present, pH, ionic strength, type
of counter-ions, etc....Of course, in the following discussion, we
thus assume the extramolecular medium to be invariable. This does
not imply that a variation of external parameters does not change the
shape of the melting curve.

If a given base pair (i) in a DNA molecule is caught in the middle of
a sequence of AT or GC base pairs, the actual melting temperature of
(i) would be low or high, respectively. Due to interactions with the
remainder of the macromolecule, the actual m.t. of (i) is defined not
only by its type, but also by its environment inside the molecule.

We are led to examine the various constraints on (i) by the remainder
of the molecule. The origin of the polyphasic melting behavior must
be related to some peculiarities in these constraints. For conve-
nience, these will be classified arbitrarily into local, intermediate,
and long-range constraints.

At a *local level*, (i) is considered to be "sandwiched" within a set
of a few base pairs. The number of base pairs in a set is taken to
be just sufficient to ensure nucleation and cooperation in its phase-
transition: in other words, if the set were cut out of the molecule
and subjected to melting conditions, its transition would be cooperative. By analogy with the behavior of oligonucleotides, for instance,
the behavior of a set upon melting can be explained by an all-or-none
scheme, as discussed by APPLEQUIST and DAMLE (1965), CRAIG et al.
(1971), and PÖRSCHKE and EIGEN (1971).

Since several sets may exist for a given (i), we define as a melting
unit (m.u.) relative to (i), μ_i, the set (or sets) of base pairs, including (i), having upon its hypothetical cut-out, the lowest melting
temperature T (μ_i).

We notice that μ_{i+1} may be identical to μ_i, or encompass one or more
different base pairs, or even be fully different from μ_i. More generally, even if (i + n) belongs to μ_i, μ_{i+n} may differ completely
from μ_i: consider, for instance, the following stretch of base pairs:

$$-GC-GC-AT-AT-AT-AT-AT-GC-GC-GC-GC-$$
$$\ \ 1\ \ \ 2\ \ \ 3\ \ \ 4\ \ \ 5\ \ \ 6\ \ \ 7\ \ \ 8\ \ \ 9\ \ 10\ \ 11$$

Assuming that 5 base pairs suffice to define a set (a purely hypothetical figure), μ_9 might encompass base pairs 3 to 9, whereas μ_7
might not include 8 and 9.

If only such local forces were acting on (i), it would melt at $T(\mu_i)$.
However, this would imply that other base pairs of μ_i, which define
other m.u.s with different m.t.s, had to melt together with (i), a
situation that may be thermodynamically unfavorable. For instance,
if $T(\mu_{i+1}) > T(\mu_i)$, the actual melting of (i) is not expected below
$T(\mu_i)$, since nucleation would then appear in μ_i. In contrast, if
$T(\mu_{i+1}) < T(\mu_i)$, nucleation for μ_i would be provided at the μ_i/μ_{i+1}
interface (or overlap), so that the melting of (i) would actually be
expected below $T(\mu_i)$.

Thus, at an *intermediate level*, constraints upon melting can exist
between overlapping or neighboring m.u.s, which may alter $T(\mu_i)$.
These constraints may be expressed quantitatively by the existence of
an interfacial energy associated with every pair of m.u.s, $E(\mu_i, \mu_j)$.
For the sake of simplicity, we will restrict the interfacial interactions between neighboring m.u.s, $E(\mu_i, \mu_{i+1})$. In terms of the nucleation parameter σ, usually introduced in the helix-coil transition
theory derived from the Ising model (CROTHERS and KALLENBACH, 1966;
FIXMAN and ZEROKA, 1968; REISS et al., 1966), this would mean that σ
is not a constant, but is a function of the position of the helix-coil
boundary in the DNA. If n is considered a variable over n = 1 to N
(the total number of base pairs in the molecule), we can define the
functions E(n) and T(n) as having the values $E(\mu_i, \mu_{i+1})$ and $T(\mu_i)$
for n = i, respectively.

The constraints described above are expressed as a modulation of T(n)
by E(n): the net effect of E(n) would be to level out to some extent
T(n) and T(n + 1).

If the mean base composition of the m.u.s of the molecule were roughly
alike, the leveling-out just described would result in a smooth melting of the DNA. Once started, melting would propagate and end when
the molecule was completely denatured. Stated otherwise, if the fluctuations of the mean base composition of the m.u.s do not exceed some
limits, which remain to be evaluated, melting would be expected to be
smooth, whatever the base-pair sequence in the various m.u.s.

However, if abrupt base composition discontinuities exist between two
neighboring m.u.s, the corresponding E(n) would be too strong to be
overcome at the temperature at which the lower melting unit denatures:
melting would stop at the m.u.s interface.

Although no direct experimental material with model polydeoxynucleo-
tides yet exists, theoretical evaluations support this conclusion.
Both MONTROLL and GOEL (1966), and POLAND and SCHERAGA (1969) have
shown that DNA of a given base composition may exhibit bimodal melting, provided the base pairs are clustered in AT- and GC-rich sequences. For instance (POLAND and SCHERAGA, 1969), DNA with x_{AT} = .5
is predicted to melt with two modes if the macromolecule in a block-
copolymer in which 12-15 ATs alternate with 12-15 GCs.

Computer-simulated melting of DNA with composition constraints yields similar results (CROTHERS, 1968; STRASSLER, 1967).

Thus, we believe that jumps in $E(n)$, induced by strong local composition variations, are mainly responsible for the polyphasic melting behavior. Melting is locally cooperative only in DNA; cooperation extends over stretches of the molecule having a fair base composition homogeneity. The melting modes reflect the existence of these stretches.

Let us finally discuss the *long-range interactions* in the molecules, which obviously cannot account for the existence of the melting modes, since they are uncoupled by the enzymatic cleavage. However, the slight differences observed between the melting curves of the pooled fragments and the intact λ DNA demonstrate that long-range interactions are indeed operative and can alter slightly the mode appearance. Several mechanisms may be responsible for the differences. First, the presence in the vicinity of μ_i of a native or denatured m.u. is expected to change the local environment of μ_i. In addition to excluded volume effects (WETMUR and DAVIDSON, 1968), it may be expected that the local ionic strength around μ_i is altered by this neighborhood, leading to an alteration of $T(\mu_i)$, especially sensitive at low ionic strength. This effect should be strongly ionic strength-dependent; its existence has been experimentally confirmed (C. REISS, unpublished). Second, $T(\mu_i)$ may also be altered because of dynamic effects coupling μ_i to other m.u.s, since upon melting of μ_i's the latter has to spin in a medium of some viscosity, which modifies its ease of unwinding (DE GENNES, 1968). Slowing down (higher $T(\mu_i)$ is observed if higher heating rates are applied to the sample (REISS and MICHEL, 1974). Dynamic effects should also be size and composition-dependent: the constraints should be heavier if μ_i belongs to a GC-rich stretch of the molecule. Finally, end effects may also account for the differences between the curves. The differences would then be localized on the mode corresponding to stretches that are hit by the enzyme, an interesting hypothesis that is presently under investigation.

D. Information Borne by a Melting Mode

It must be emphasized that a melting mode as seen on a denaturation curve of DNA is only operationally defined. It cannot be ascertained that only *one* segment of the macromolecule contributes to a given mode. It may be that investigations with even more sensitive equipment would reveal a splitting of yet unresolved modes. Unfortunately, the theoretical limit of the melting mode width is not clear, since contradictory estimations have been published (CROTHERS et al., 1965; EICHINGER and FIXMAN, 1970). This is, however, not a hopeless situation, because the partial unwinding-rewinding experiments that will be described later, together with melting of small, homologous fragments of DNA, can give some clues as to the mode "purity."

Another problem concerns the mathematical form that suitably represents the melting modes. Arguments have been developed that indicate that melting modes should be non-symmetric, building up slowly on the low-temperature side (where nucleation is to appear within the segment) but decreasing abruptly on the high temperature side (CROTHERS, 1968). Since our experiments are performed at equilibrium, this dissymmetric effect is minimized; we assume, for the sake of simplicity, that the melting of a segment is a stochastic process, so that the corresponding melting mode can be assumed to be gaussian in shape.

Transformation of the derivative melting profile into gaussian distributions can be efficiently handled by a computer. Then, every mode can be characterized by its surface, S, the position of its maximum, Tm, and its width at half maximum, ΔT. Transformation is performed in two steps: recognition of the modes through an algorithm providing numerical filtering, yielding Tm, followed by a random minimization treatment, which generates ΔT and S, so that the sum curve of the gaussian distribution fits the experimental curve.

It is clear that the number of gaussian distributions obtained by numerical filtering depends on the width of the filter band, which in turn depends on the temperature step of the experiment. As a rule, the filter band width is chosen so that at least eight measurements are required for the definition of a mode. For instance, if the experiment is conducted with 0.05°C steps, the smallest mode considered meaningful has a 0.4°C width.

The meanings of the gaussian parameters are obvious: Tm indicates the mean base composition of the mode, provided the Marmur-Doty relationship still applies. The careful investigations by OWEN et al.,(1969) on the validity of this relationship have been performed for whole DNA only, so that extrapolation to intramolecular parts is questionable. In order to establish firmly this point, composition studies of isolated, well-characterized fragments of DNA should be performed by some independent method, for instance by buoyant density measurements. In the absence of such experiments (which are on the schedule), we are left with indices showing that the Marmur-Doty relationship applies roughly, at least, at a local level. CHAMPOUX and HOGNESS (1972) established a composition map of λ fragments; on the other hand, a map derived from thermal transition experiments (BRAM et al., 1974) has been drawn. Although no comparison on an absolute scale can be made owing to the fact that the latter map assumed the Marmur-Doty relationship, the shapes of both maps are similar, so that one can at least assume that at a local level in the DNA, the variations of mean composition and melting temperature are correlated.

The surface S of the mode is proportional to the length of the stretch (or stretches) contributing to the mode. Normalization of the total melting curve surface and additional correction due to the mean composition of the mode have to be performed (AT pairs contribute about twice as much to hypochromicity as GC pairs at 260 nm; HIRSCHMAN and FELSENFELD, 1966).

Finally, ΔT is a measure of the standard deviation of the base composition from its mean value; it reflects the composition fluctuation in the stretch (or stretches). We propose to call *thermosome* a stretch of the DNA molecule giving rise to a gaussian mode in the derivative melting curve.

E. Biologic Significance of the Thermosomes

Several questions arise:

1. How can the local base composition constraints be reconciled with the genetic code?
2. How are these constraints affected by the endogenous media surrounding the DNA in vivo?
3. Is there any correlation between the stretches and the organization of the genetic function of DNA?
4. Are these findings restricted to phage or virus DNA?

1. The degeneracy of the code allows for large composition fluctuations of the DNA, coding for a set of amino acids. For instance, it is possible to build sequences of DNA containing all the codons for the 21 amino acids, with composition ranging between 68% AT and 63% GC. However, every percentage between these two extremes cannot necessarily be fitted. This observation may be related to the appearance of discrete melting modes.
2. Although precise structural data for AT- or GC-rich sequences are lacking, BRAM (1972) has shown that DNAs can exhibit several tridimensional structures, according to the mean base compositions. Assuming that these observations can be extrapolated to local parts of the DNA, one could conceive that a thermosome characterized by a specific mean base composition has a specific tridimensional structure. A thermosome may have some particular double helical structure, which of course could have an important biologic implication, for instance, in DNA-ligand recognition and specificity.
3. To decide whether a correlation exists between the thermosome and the organization of the DNA, the former must, of course, be localized.

A mapping procedure is described in the next paragraph. In one instance at least, which is described in detail, it seems that the thermosome may bear some correlation with the gene.

4. The multimodal melting behavior is not restricted to virus or phage DNA. Mitochondrial DNA from yeast (MICHEL et al., 1974), chloroplast (PIVEC et al., 1972), and calf thymus DNA (WELLS and SAGER, 1971) also show complex melting curves. However, except in particular cases (satellite DNA, yeast, mutant mitochondrial DNA), the presence of discrete modes is not obvious. We believe that this results from a blurring effect, due to the existence of the many modes that are present in these highly complex DNAs. Isolated fragments obtained from the DNA by the endonuclease cleavage technique are expected to yield melting curves with fine structure.

F. The Thermosome Map

The denaturation profile as such can obviously not yield any information on the relative position of thermosomes, since the positions of the corresponding melting modes depend mainly on the mean base composition. However, a set of partial denaturation experiments followed

by renaturation under well-defined experimental conditions can yield
an unambiguous topology of the thermosomes. This will now be demonstrated qualitatively, with the aid of a simplified example.

Let us first define the experimental conditions under which the experiments are conducted. We define as "reassociation" the phenomenon
that brings back into register two fully separated strands of DNA,
and as "renaturation" the same phenomenon for two strands that still
bear native parts. The two phenomena can be distinguished by the
speeds of rewinding, at nondenaturing temperatures: in high salt, the
repulsion between the two strands is shielded by the large amount of
counterions, so that nucleation can easily take place between two
complementary but fully denatured strands (rate constant depending on
strand concentration, ionic strength, temperature...). Once nucleation is provided, rezippering of the two strands can proceed, just
as for two strands with double helical parts still present. Nucleation times depend on DNA complexity, concentration, ionic strength,
etc..., but are typically in the range of hours for phage DNA in
$\sim 10^{-1}$ M Na$^+$. However, if the ionic strength is lowered to below
$\sim 10^{-2}$ M Na$^+$, nucleation takes days, so that it does not occur significantly over the time of a mapping experiment as we will describe
it. Thus, reassociation is eliminated in $\sim 10^{-2}$ M Na$^+$, whereas renaturation can occur, owing to the presence of rezippering nuclei in
the only partially denatured material.

We assume that the definitions of renaturation and reassociation still
hold for thermosomes. The two strands of a thermosome may be fully
separated, with no native part even at the ends of the thermosome under consideration; at low ionic strength, this thermosome is not sup-
posed to reassociate upon lowering the temperature beyond the melting
range. If, however, at least one native part is present at the thermosome's boundaries, i.e., a neighboring thermosome is native, then
renaturation can occur, starting at this boundary. Obviously, if
compared to the kinetics of reassociation of fully separated strands,
the kinetics of reassociation of thermosome strands with no native
boundary, but belonging to a still partially double-stranded DNA, will
be speeded up; however, this is not likely to occur to the extent that
reassociation occurs during a P.U.R. experiment, especially if ionic
strength is lowered beyond 10^{-2} M Na$^+$.

G. Mapping of Thermosomes through Partial Unwinding-Rewinding (P.U.R.) Experiments

The method has been developed together with F. MICHEL. A detailed
description can be found elsewhere (MICHEL, 1974; REISS, in prep.).
We will just outline here the main features of the method.

To do this, we consider the problem in the reverse: How does the
thermosome position influence the partial denaturation-renaturation
pattern? We assume that the ionic strength of the medium is beyond
10^{-2} M Na$^+$. We are dealing with a hypothetical DNA, built up with
five thermosomes: two, (A) and (B), are AT-rich; three, (1), (2), (3)
are GC-rich, so that the melting profile of the molecules looks like

<u>Fig. 12</u>. Behavior upon melting and rena-
turation of thermosomes A and B when T_B
is turn-back temperature (dotted line
corresponds to renaturation profile)

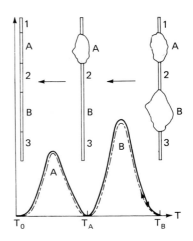

that shown in Figure 11. The five thermosomes are arranged as shown
in the legend of Figure 11.

Upon melting the DNA to temperature T_A or T_B, thermosomes A, or respectively, A and B denature (Fig. 12). Lowering the temperature
back to T_O, both A and B will *renature*, because nucleation is provided
at their boundaries with neighboring thermosomes that are still native:
between T_B and T_O, for instance, the hypochromicity variation should
have the same absolute value as the corresponding variation upon melting from T_O to T_B; melting and rewinding profiles are identical. If
now the temperature is raised to T_1, (A), (B), and (1) are denatured.
Upon lowering the temperature from T_1 to T_A, (1) should renature, but
it cannot, because no nucleation is available, owing to the denatured
state of (A); no hypochromicity recovery is obtained between T_1 and
T_B (Fig. 13); between T_B and T_A, (B) is renaturing as previously,
starting at the (B) - (2) or (B) - (3) interface; between T_A and T_O,
(A) is rewinding since nucleation is provided at the (A) - (2) interface. Now, once (A) is renatured, (1) can renature also: between T_A
and T_O, the hypochromicity recovery due to thermosomes (A) and (1) is
observed (Fig. 13). This means that (1) and (A) are neighbors. Melting to T_2, followed by renaturation to T_1, will similarly demonstrate
the (2) - (B) connection (Fig. 14).

To complete the mapping, additional information can be gathered from
strand breaks that occur spontaneously in the sample. It is well
known that DNA is a sturdy molecule in the native state, but that
this is much less so for single-stranded, denatured DNA. Actually,
breaks are known to occur in DNA strands about one thousand times
more often in the single-strand state than in the native one (MARMUR
and DOTY, 1962). The breaks seem mainly due to depurination, but let
us assume that they occur at random, so that the number of breaks increase with the time of single-strandedness according to a Poisson
law.

Let us now focus attention on thermosome (A). During P.U.R. experiments to T_A and T_B, (A) remains single-stranded for some well-known

24

Fig. 13. Behavior upon melting and
renaturation of thermosomes A, B,
and (1) when T_1 is turn-back temperature (dotted line corresponds
to renaturation profile)

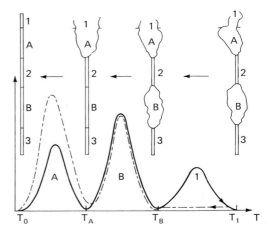

Fig. 14. Behavior upon
melting and renatura-
tion of thermosomes A,
B, and (1) and (2) when
T_2 is turn-back temperature (dotted line cor-
responds to renatura-
tion profile)

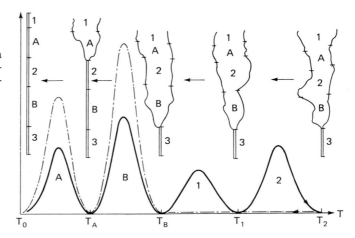

time, which implies that the number of single-strand cuts, which can
be computed, will occur in (A). Since every cut in a given strand-
except the first will decrease the amount of hypochromicity recovery
when the temperature is lowered to T_0, the assumed randomness of
single-strand breaks allows the computation of the hypochromicity
loss of (A) during P.U.R. experiments to T_A and T_B.

Now, when melting to T_1 is performed, thermosome (1) becomes single-
stranded. If (1) is at the end of the DNA as pictured, the strands
of (1) linked to strands of (A), having been hit by one (or several)
break(s), will leave the DNA, and will not subsequently reassociate.
The result is a massive loss of hypochromicity recovery between T_A
and T_0, where (1) + (A) should renature. Quantitatively, the loss
will be much higher than that which could be computed, if only single-
strand breaks had occurred in (1). In contrast, if the relative positions of (1) and (A) were reversed, the breaks in (A) would not
have had any effect on the hypochromicity recovery of (1), which would
only have been a function of the time (1) remains denatured.

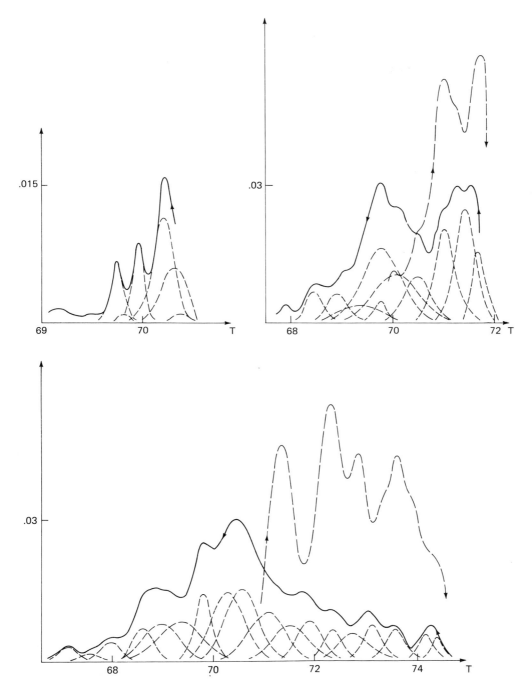

<u>Fig. 15a-c.</u> Three typical P.U.R. experiments for left 40% of λ plac DNA

A full, quantitative treatment of the principle briefly outlined will
be published elsewhere (MICHEL, 1974; REISS, in press).

An illustration of the mapping method is given with the mapping of the
left hand, 40% genome of λ plac, which has been prepared by Eco R_1
cleavage by P. TIOLLAIS and A. FRITSCH (Pasteur). The experiment was
conducted as follows: a complete melting curve was obtained. The
melting modes were then detected, and a set of 15 P.U.R. experiments
performed, with turn back temperatures located just above each.of the
15 higher melting modes. A fresh sample, taken from the stock, was
introduced into the cuvette-holder after every P.U.R. experiment, and
subsequently measured together with the samples already under investigation. A given sample was thus subjected to 6 P.U.R. runs, so that
hypochromicity losses could be monitored accurately. Three typical
P.U.R. runs are shown in Fig. 15, with the full melting profile of the
fragment; Fig. 16 shows the map, which has been drawn according to the
rules just outlined. A full account of the experiment will appear
shortly (REISS et al., in prep.).

Surprisingly enough, some correlations seem to exist between the
thermosome map and the genetic map (PARKINSON, 1968) shown in the
same figure. Whether this is fortuitous or not will soon be investigated.

III. Application of the Method to the Study of DNA-Containing Complexes

Thermal transition spectroscopy proved to be a valuable tool for submolecular investigation of DNA. Concerning the study of complexes in

Fig. 16. Map of left 40% of λ
plac DNA. Caption shows genetic
map of this part of λ DNA, re-
drawn from REISS et al., (in
prep.) *Ordinate*: melting temperature; *abscissa*: % length of
genome

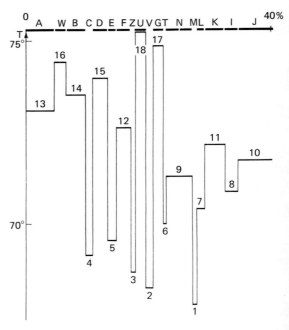

which DNA is involved, some unique results can be obtained by the
method. The basic idea is simple: when a complexing particle binds
to a DNA element, the corresponding melting mode is perturbed, since
the interaction stabilizes or destabilizes the element. Thermal Transition Spectroscopy can yield the following information, some of which
is not accessible by any other known method:

1. *The interaction type* (stabilization, destabilization) can be estimated by comparing the melting profiles of pure DNA and complexed
   DNA.
2. *The interaction energy* can be measured by the temperature shift
   between the melting profiles of pure and complexed DNA.
3. *The local base composition* of the DNA at the attachment site can be
   identified through the melting temperature of the thermosome(s)
   affected by the complexation.
4. *The mapping of the attachment site* can be obtained if mapping of
   the thermosome has been achieved (this is possible for DNA of less
   than 10^7 base pairs).
5. *The mechanism of the interaction* can be studied by plotting the
   rate of mode modification as a function of complexing agent concentration. Sigmoidal curves, for instance, may indicate that the
   mechanism is cooperative, that is, that the presence of a bound

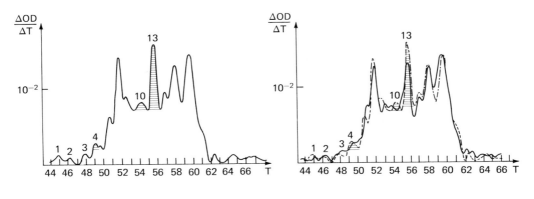

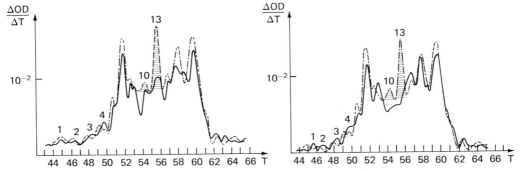

Fig. 17a-d. Melting of λ´ plac DNA and λ plac DNA calf thymus histone fraction F_1 complexes in SSC x 10^{-2} (a) 0% F_1; (b) 1% F_1; (c) 5% F_1; (d) 10%
F_1 (% weight protein/weight DNA). Missing hypochromicity in complexes
appears above 80°C. See BRAM et al. (1974) for details. (Reprinted by
permission of the publisher of Biochimie)

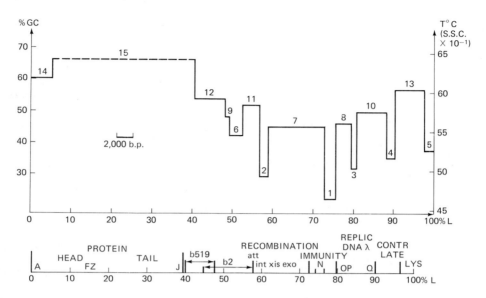

Fig. 18. Rough map of λ plac DNA; *ordinate*: % GC (*left*) or melting temperature in SSC x 10⁻² (*right*); *abscissa*: % length of DNA molecule

complexing agent facilitates the binding of the next on the same molecule.

The study of DNA-containing complexes will be illustrated by a published example.

The example deals with the specific interaction of histone fraction F_1 from calf thymus and λ plac DNA. Obviously, F_1 interacts specifically with the thermosomes corresponding to peaks 4, 10, and 13 of the melting profile (Fig. 17) (BRAM et al., 1974). A rough map of the thermosome has been established (Fig. 18). It appears that thermosomes 4, 10, and 13 are located in the right arm of the λ genome, in the neighborhood of the N thermosomes gene. This specificity is perhaps fortuitous; the example shows, however, the potentialities of the method.

Before concluding, we would like to mention that the method has just been applied to the study of proteins and RNAs, with promising results that will be published soon.

Upon considering the results already obtained, it appears that the old thermal denaturation technique can be supplemented by a few simple improvements. It turns out that it can then be a powerful tool, which is able to yield information on biologic macromolecules at a submolecular level. This information is much needed in molecular biology.

Acknowledgments

The progress achieved during this work depended heavily on the DNA
samples given to the authors. Many thanks are due to the research
groups, which carefully cultured the microorganisms and extracted the
DNA. Support and encouragement by Dr. SLONIMSKI at the beginning of
this work is acknowledged. Mr. DELHAYE (DATEC) provided skillful assistance in designing and building the electronic part of the device.
Dr. REULE (Carl ZEISS) gave valuable advice with the spectrophotometer
interfacing. Computer operation assistance by Patrick GAUTIER is also
acknowledged.

References

APPLEQUIST, J., DAMLE, V.: Thermodynamics of the helix-coil equilibrium in oligoadenylic acid from hypochromicity studies. J. Amer.
Chem. Soc. 87, 1450 (1965).

BERNARDI, G., FAURES, M., PIPERNO, G., SLONIMSKI, P.P.: Mitochondrial
DNA's from respiratory-sufficient and cytoplasmic respiratory-
deficient mutuant yeast. J. Mol. Biol. 48, 23 (1970).

BRAM, S.: The polymorphism of natural DNA. Biochem. Biophys. Res.
Commun. 48, 1088 (1972).

BRAM, S., BUTLER-BROWN, G., BRADBURY, E.M., BALDWIN, J., REISS, C.,
IBEL, K.: Chromatin neutron and X-ray diffraction studies, and high
resolution melting of DNA-histone complexes. Biochimie 56, 987
(1974).

CHAMPOUX, J.J., HOGNESS, D.S.: The topography of lambda DNA: poly-
riboguanilic acid binding sites and base composition. J. Mol. Biol.
71, 383 (1972).

CRAIG, M.E., CROTHERS, D.M., DOTY, P.: Relaxation kinetics of dimer
formation by self complementary oligonucleotides. J. Mol. Biol. 62,
383 (1971).

CROTHERS, D.M.: Calculation of melting curves for DNA. Biopolymers 6,
1391 (1968).

CROTHERS, D.M., KALLENBACH, N.R.: On the helix-coil transition in
heterogeneous polymers. J. Chem. Phys. 45, 917 (1966).

CROTHERS, D.M., KALLENBACH, N.R., ZIMM, B.H.: The melting transition
of low molecular weight DNA: theory and experiment. J. Mol. Biol.
11, 802 (1965).

DE GENNES, P.G.: Regimes transitoires dans une transition complete
helice pelote statistique. J. Chem. Phys. 48, 962 (1968).

EICHINGER, B., FIXMAN, M.: Helix-coil transition in heterogeneous
chains. II. DNA model. Biopolymers 9, 205 (1970).

FALKOW, S., COWIE, D.B.: Intramolecular heterogeneity of the deoxy-
ribonucleic acid of temperate bacteriophages. J. Bacteriol. 96, 777
(1968).

FIXMAN, M., ZEROKA, D.: Helix-coil transition in heterogeneous chains.
I. Protein model. J. Chem. Phys. 48, 5223 (1968).

HIRSCHMAN, S.Z., FELSENFELD, G.: Determination of DNA composition and
concentration by spectral analysis. J. Mol. Biol. 16, 347 (1966).

INMAN, R.B.: A denaturation map of the λ phage DNA molecule determined by electron micrography. J. Mol. Biol. 18, 464 (1966).

LEHMAN, G.W.: Thermal properties of linearly associated systems with random elements: helix-coil transition in DNA. In: Conference on Statistical Mechanics and Thermodynamics (Ed. T.A. BAK). New York: W.A. Benjamin 1967.

MARMUR, J., DOTY, P.: Determination of the base composition of DNA from its thermal denaturation temperature. J. Mol. Biol. 5, 109 (1962).

MICHEL, F.: Hysteresis and partial irreversibility of denaturation of DNA as a means of investigating the topology of base distribution constraints: application to a yeast ρ^-(petite) mitochondrial DNA. J. Mol. Biol. 89, 305 (1974).

MICHEL, F., LAZOWSKA, J., FAYE, G., FUKUHARA, H., SLONIMSKI, P.P.: Physical and genetic organization of petite and grande yeast mito-chondrial DNA. III. High resolution melting and reassociation studies. J. Mol. Biol. 85, 411 (1974).

MONTROLL, E.W., GOEL, N.S.: Denaturation and renaturation of DNA. I. Equilibrium statistics of copolymeric DNA. Biopolymers 4, 855 (1966).

OWEN, R.J., HILL, L.R., LAPAGE, S.P.: Determination of DNA base composition from melting profiles in dilute buffers. Biopolymers 7, 503 (1969).

PARKINSON, J.S.: Genetics of the left arm of the chromosome of bac-teriophage λ. Genetics 59, 311 (1968).

PIVEC, L., STOKROVA, J., SORMOVA, Z.: Plurimodal heterogeneity of base composition of calf thymus DNA. Biochim. Biophys. Acta 272, 179 (1972).

POLAND, D., SCHERAGA, H.A.: The Lifson-Allegra theories of the helix-coil transition for random copolymers: comparison with exact results and extension. Biopolymers 7, 887 (1969).

PÖRSCHKE, D., EIGEN, M.: Co-operative non-enzymic base recognition. III. Kinetics of the helix-coil transition of the olig U-oligo A system, and of oligo A alone at acidic pH. J. Mol. Biol. 62, 361 (1971).

REISS, C.: Thermal transition spectroscopy of nucleic acids. IV. Sequence mapping. Submitted for publication.

REISS, C., MICHEL, F.: An apparatus for studying the thermal transi-tion of nucleic acids at high resolution. Anal. Biochem. 62, 499 (1974).

REISS, C., TIOLLAIS, P., FRITSCH, A.: Thermal transition spectroscopy of nucleic acids. V. Composition mapping of the left arm of λ DNA. (Manuscript in preparation).

REISS, H., McQUARRIC, D.A., McTAGUE, J.P., COHEN, E.R.: On the melting of copolymeric DNA. J. Chem. Phys. 44, 4567 (1966).

STRASSLER, S.: Theory of the helix-coil transition in DNA considered as a periodic copolymer. J. Chem. Phys. 46, 1037 (1967).

TIOLLAIS, P., RAMBACH, A., BUC, H.: Large scale purification of bac-terial genes. FEBS Letters 48, 96 (1974).

WELLS, R., SAGER, R.: Denaturation and renaturation kinetics of chloroplast DNA from Chlamydomonas reinhardi. J. Mol. Biol. 58, 611 (1971).

WETMUR, J.G., DAVIDSON, N.: Kinetics of renaturation of DNA. J. Mol. Biol. 31, 349 (1968).

Interactions of Drugs with Liver Microsomes

Ingeborg Schuster

I. Introduction

The history of microsomes dates back to 1938 when CLAUDE first detected a new cell fraction in mammalian cells. For these particles,
which could be isolated by differential centrifugation and afterward
observed in a dark field microscope "as fine granules," he proposed
the term microsomes (CLAUDE, 1943). With the newly evolving technique of electron microscopy it was then proved that microsomes resembled the isolated form of endoplasmic reticulum (CLAUDE et al.,
1947; PALADE and SIEKEVITZ, 1956).

After these reticular membranes had been obtained in a purified form,
they were shown to be the sites of a large number of different metabolic activities. Besides their important role in the synthesis of
protein (CAMPBELL and LAWFORD, 1968), of fatty acids (LANDRISCINA et
al., 1970) and of phospholipids (DE KRUIFF et al., 1970), microsomes
contain extensive systems designed to convert lipophilic compounds to
more water soluble products. Substrates of these enzyme systems are
both endogenous, such as steroids and certain amino acids, and exogenous substances like drugs, carcinogens, and other environmental
chemicals (for review see: SIEBERT, 1968).

A wide variety of oxidation reactions occur in microsomes, which can
be visualized as different kinds of hydroxylation reactions. Recognition that the hemoprotein, cytochrome P450, is the key enzyme in
most of these hydroxylation processes (ORRENIUS and ERNSTER, 1964)
has focused the main interest on the properties of this protein,
which is intimately associated with microsomal membranes.

Besides the conversion of drugs by an enzyme system containing cytochrome P450, microsomes also metabolize drugs by other mechanisms
such as cleavage of ester bonds and formation of glucuronides.

Microsomal interactions with drugs are not only confined to processes
that lead to metabolic changes of the substances. From the fact that
microsomes consist of lipoprotein membranes, an abundance of sites
capable of interactions with lipophilic drugs can be considered.
However, there is still little information on that kind of drug interaction. The binding of drugs to these "nonaction" sites also has
important pharmacologic implications. In their passage through the
liver, blood levels of drugs decrease due to high accumulation in

endoplasmic structures. This enrichment may then influence-directly or indirectly by membrane alterations-microsomal activities.

It is the aim of this article to present a summary of the ways in which drugs can interact with liver microsomes. Liver was chosen since it is generally accepted to be in most cases the major site of drug biotransformation and has therefore stimulated much work on microsomal drug metabolism.

First, this survey will deal with the characterization of biochemical entities in microsomes, and then their possible involvement in drug interactions will be discussed. However, the scope will be restricted to intrinsic constituents of the membranes and will not include drug effects upon the activities of attached ribosomes, which would require a separate article. (For detailed treatment of drug-ribosome-interactions, the reader is referred to PESTKA, 1971 and VAZQUEZ et al., 1975.)

A large part will necessarily be devoted to the central role of the cytochrome P450-containing enzyme system in binding and subsequent metabolism of drugs.

II. Structural and Chemical Organization of Liver Microsomes

A. Morphologic Aspects

1. Intact Endoplasmic Reticulum

In the fine structure of the cell, the endoplasmic reticulum resembles a network of lipoprotein membranes that furrow the cytoplasm. Electron microscopy was the technique principally used to elucidate the morphologic features of this cell constituent (CLAUDE et al., 1947; PALADE and SIEKEVITZ, 1956), which consists of dynamic structures of channels, tubules and cisternae, variable in size and substructure, which enclose spaces and thereby separate them from cytoplasma. Reticular membranes form connections between different cellular components, e.g., between nuclear envelope and outer membrane, and thus provide channels for the transport of various compounds.

Two distinct types of reticulum can be observed (PALADE and SIEKEVITZ, 1956). In the rough or granular reticulum, the membranes are studded with ribosomes. These particles are reversibly attached to the outside of the membranes through the large subunit (SABATINI et al., 1966) by ionic forces (SHIRES et al., 1975). Membranes that are devoid of ribosomes constitute the smooth reticulum. Both types of membranes appear to be continuously linked together but resemble different systems with regard to their distribution and organization in the cytoplasmic matrix (CLAUDE, 1969), their biochemical constitution (GRAM et al., 1971; MOULÉ, 1968), and hence, their physiologic functions.

As was demonstrated by DALLNER et al. (1966), the constituents of smooth membranes are synthesized in the rough reticulum and subsequently transferred to the newly emerging smooth regions. In both types of membranes the components are not randomly distributed, but distinct segments, specialized for limited functions, were observed (SVENSSON et al., 1972).

2. *Formation of Microsomes*

On disrupting the cell, the extended reticular membranes are transformed into spheres with diameters of about 200 mμ called microsomes (CLAUDE, 1948). This vesiculation obviously represents a pinching-off process without leakage of the contents of endoplasmic cisternae (CLAUDE, 1969). The breakage of the membranes probably occurs at defined points of the system (SVENSSON et al., 1972).

The resulting vesicles respond osmotically to variations in the composition and salt concentration of the media (CLAUDE, 1969). Uncharged molecules up to molecular weight of more than 600 penetrate. However, the membranes are impermeable to small anions with molecular weights of as low as 90 (NILSSON et al., 1971).

Microsomes retain many of the functional properties of the intact reticulum and serve as a representative model for biophysical and biochemical studies.

B. Chemical Constitution

The proportion of endoplasmic membranes in the total cell is variable and depends on the cell type. In mammalian liver, the microsomal fraction amounts to as much as 25% of total cell dry weight. The rough chemical constitution of microsomes is shown in Table 1.

1. *Microsomal RNA*

Most of the RNA, localized in the ribosomes, consists of two components sedimenting at 28S and 18S (MOULÉ, 1968). However, it was shown that smooth membranes without any visible ribosome also contain RNA (MOULÉ, 1968). This RNA differs from the ribosomal one in base composition (more guanine), association with lipids, and metabolic activity (MOULÉ, 1968).

Table 1. Biochemical classes of compounds in liver microsomes (BIELKA, 1969; KHANDWALA and KASPER, 1971)

Class	Compound x 100 total microsomal fraction	Compound in microsomes x 100 compound in total liver
Protein	56	20
Lipid	31	50
RNA	9	60
Carbohydrate	2.5	
% of total weight	100	25

Table 2. Composition of the lipid fraction in rat liver microsomes (SCHULZE and STAUDINGER, 1971; GLAUMANN, 1970)

Constituents	Percentage of lipid fraction (a) of total lipid (b) of phospholipid
Phospholipids	80-85 (a)
lecithins	55 (b)
cephalins	22 (b)
phosphatidyl inositol	9 (b)
phosphatidyl serine	7 (b)
sphingomyelin	5 (b)
phosphatidic acids and lysophosphatides	2 (b)
Cholesterol	5-10 (a)
Triacylglycerols, free fatty acids, and cholesterol esters	6 (a)

2. Microsomal Lipids

Among all cell organelles, microsomes have by far the highest lipid content. Phospholipids constitute about 80-85% of the total lipid (SCHULZE and STAUDINGER, 1971). As is a general feature of inner cell membranes, the cholesterol content of microsomes is low, amounting to about 5-10% of the total lipid (GLAUMANN, 1970; GRAM et al., 1971; SCHULZE and STAUDINGER, 1971). Glycolipids are naturally absent from microsomal membranes (ROUSER et al., 1968).

Table 2 summarizes the constituents of the lipid fraction from rat liver microsomes.

Rough and smooth surfaced reticula from the same organ do not differ significantly in their phospholipid composition. On comparing the lipid constitutents of any organ in different animal species, no note-worthy difference is found (ROUSER et al., 1968). However, microsomes from different organs of one species show essential variations (ROUSER et al., 1968).

3. Microsomal Proteins

a. General

With the exception of a few enzyme systems that have been isolated and (partially) purified, microsomal proteins are still a poorly charac-terized mixture of heterogeneous structures.

An appreciable part of them were found to be glycoproteins (KAWASAKI and YAMASHINA, 1973). The total carbohydrate content (2.5% of micro-somal dry weight) consists of about 60% neutral sugars, 36% hexosamine and 4% sialic acid (KHANDWALA and KASPER, 1971). Rough and smooth membranes differ insofar as neutral components are major components in rough reticulum, whereas acid glycopeptides predominate in smooth membranes (KAWASAKI and YAMASHINA, 1973). In relation to protein content, rough reticulum contains only half as much amino sugars (about 38 μm/mg protein) as smooth reticulum.

KHANDWALA and KASPER (1971) investigated the accessibility of different amino acid side chains in the native microsomal membrane. About 85% of the tyrosyl residues appeared to be easily accessible, whereas tryptophan was by and large inert to succinylating agents, and about 40-50% of lysyl residues were also hidden in the interior of the membrane structure. This suggests that tryptophan and lysine may play an important role in the maintenance of membrane structure.

According to their association with the membrane, microsomal proteins can be roughly divided into two types: "foreign" proteins, which are either directly or indirectly attached to the membrane in a relatively loose association, and "real" microsomal proteins, which are integral constituents of the membranes. Most microsomal enzyme systems belong to the latter group.

b. *Foreign Proteins*

Ribosomal proteins may be assigned to this group of proteins. The heterogeneous mixture, which contributes about 10% to total microsomal protein, has been resolved by gel electrophoresis into individual components that have been further characterized (BIELKA, 1969). However, they will not be dealt with in detail in this article.

Bound ribosomes are the sites where a multitude of proteins are synthesized, designed either for "export" or as new constituents of microsomal membranes themselves: Besides the nascent polypeptide chains that are attached to the polysomes, newly produced proteins are associated with the reticulum during their passage over smooth membranes to the Golgi apparatus. These proteins are also found within the microsomal vesicles. About 1% of total microsomal protein is found to be albumin (PETERS et al., 1971) and about 0.5% is contributed by cytochrome c, which is also manufactured by bound polysomes (CAMPBELL and LAWFORD, 1968). Associated glycoproteins from plasma also form a significant proportion.

The amount of foreign proteins was estimated to be about one half of total microsomal protein (DALLNER, 1969; GLAUMANN and JAKOBSSON, 1969). These nonmembranous proteins consist of about 50% protein adsorbed from cytoplasma, 30% intraluminal proteins, and about 20% ribosomal proteins.

c. *Microsomal Enzymes*

Electron Transport Chains. An essential role in the metabolism of various endogenous and foreign lipophilic compounds is played by several electron transport chains, which are intimately associated with microsomal membranes.

Of particular importance herein is the system containing the hemoprotein cytochrome P450 as its terminal oxidase and requiring NADPH and molecular oxygen. This system is exceptionally versatile with respect to its substrates and, moreover, is rich in microsomes. The content of cytochrome P450 ranges from less than 4% of total micro-somal protein to about 20% in phenobarbital-treated rabbits (WICKRAMASINGHE, 1975; ESTABROOK et al., 1971).

In the electron transport chain, cytochrome P450 is preceded by its NADPH-dependent reductase. This flavoprotein with a molecular weight of 80,000 (DIGNAM and STROBEL, 1975), containing both FMN and FAD (VAN DER HOEVEN and COON, 1974), can be calculated to total about 0.5% of microsomal protein.

A second electron transport chain containing the hemoprotein cyto-chrome b5 as an intermediate electron carrier and a NADH-dependent flavoprotein, capable of reducing the cytochrome, also contributes considerably to microsomal proteins. Using their spectral properties and the molecular weights of 16,700 for cytochrome b5 and 43,000 for the reductase (SPATZ and STRITTMATTER, 1971, 1973), the contents of both proteins can be calculated to be about 1% for the cytochrome and 0.65% for its reductase (ESTABROOK and COHEN, 1969).

A further mixed-function oxidase has been purified by ZIEGLER and MITCHELL (1972) and shown to be a flavoprotein. This enzyme catalyzes the N-oxidation of secondary and tertiary amines and is also capable of S-oxidation.

   Further Enzymes. Besides their crucial importance in the oxidation of lipophilic compounds, liver microsomes possess a further variety of enzymes with activities specific to the organ. With regard to drug metabolism, three additional enzymes must primarily be considered: The conjugating enzyme uridine diphosphate glucuronide-glucuronyl-transferase plays an essential role in the terminal metabolism of drugs. This protein is firmly associated with microsomal membranes, and its catalytic activity clearly depends on the fluidity of the surrounding lipid phase (ELETR et al., 1973). Recently, PUUKKA et al. (1975) succeeded in concentrating the enzyme 100-fold with respect to its specific microsomal content.

Though microsomes are not the only sites in which unspecific esterases are localized, their content of these enzymes is rather high. Several multiple forms of esterase occur fairly loosely attached to the mem-branes, amounting to a total of 7% of microsomal protein (KUNERT and HEYMANN, 1975).

Epoxides that can be oxidatively formed from polycyclic hydrocarbons (see Sect.V.B.) are converted by epoxide hydrase to much less reactive vicinal diols. This enzyme is exclusively located in microsomal mem-branes, obviously near cytochrome P450 mono-oxygenase (OESCH, 1972). OESCH proposes a coupled mono-oxygenase-epoxide hydrase multienzyme complex that shortens the lifetime of intermediate products before they react with nucleophilic moieties of cell constituents.

Microsomes were shown to be closely involved in the metabolism of fatty acids and phospholipids:

They are able both to elongate fatty acid chains and to synthesize fatty acids de novo. Obviously, ATP is a regulatory factor in these processes, since in the presence of much ATP the former operates to the exclusion of the latter, whereas in the absence of ATP de novo synthesis starts (LANDRISCINA et al., 1970).

Microsomal acyl transferase links the resulting fatty acids to L-glycerol-3-phosphate, forming a 1,2-diacyl-product that is a key compound in lipid biosynthesis. The enzymes choline phosphotransferase, ethanolaminotransferase, and diacyl glycerol-transferase are located only in microsomes and catalyze the formation of lecithins, cephalins, and triacylglycerols (DE KRUIFF et al., 1970).

The individual steps of cholesterol synthesis occur in different cell structures. The enzyme system catalyzing the conversion of squalene to 2,3 oxidosqualene and a further enzyme 2,3 oxidosqualene cyclase, which forms lanosterol in a consecutive step (YAMAMOTO and BLOCH, 1969), are localized in microsomes.

Glucose-6-phosphatase provides a link to carbohydrate pathways. This specific microsomal protein, which is essential for the formation of blood glucose, is fixed to the membranes so tightly and depends so much on its lipid surroundings that attempts to obtain a purified preparation have failed to date (GARLAND et al., 1974).

The hydrolysis of nucleotide triphosphates of adenosine, guanosine, uridine, cytidine, and inosine to the corresponding diphosphates and inorganic phosphate is catalyzed by a triphosphatase that is firmly bound to microsomal membranes; however, the resulting products are cleaved by a diphosphatase associated with the membrane in a relatively loose linkage that is obviously not essential for the enzymatic action (ERNSTER and JONES, 1962).

III. Interaction of Drugs with the Microsomal Lipid Fraction

A. General

In 1959, GAUDETTE and BRODIE proposed that the drug-metabolizing enzymes are protected by a lipoidal barrier that is penetrated only by fat-soluble substances. They also demonstrated a relationship between the lipophilicity of compounds—judged by their chloroform-buffer partition coefficient—and the ease of metabolism.

Meanwhile, the necessity of phospholipids for the function of various microsomal enzymes has been established (ELETR et al., 1973; CHAPLIN and MANNERING, 1970). From spin label experiments STIER and SACKMANN

(1973) proposed that mosaic-like structures exist in the membranes: the bulk of the lipid exists in a rather fluid state, whereas a relatively rigid phospholipid halo is formed around the enzyme systems.

The molecular arrangement of integral membrane enzymes within the supporting lipid bilayer and the manner in which they interact with their environment are still unresolved problems. However, the fluidity of lipids has been demonstrated to be essential for the function of several integral membrane proteins: complexes of purified ATPase with defined lipids showed a complete inhibition of ATPase activity when the associated lipids were in a crystalline state (METCALFE, 1975). A variety of carrier-mediated processes were also shown to depend strictly on lipid fluidity (TRÄUBLE and EIBL, 1975).

Lipophilic drugs will interact with the lipid part of the membrane. Enrichment can change the properties of the membrane and may influence the activity of the imbedded enzymes either directly, by binding of drug in the nearest neighborhood of lipid-dependent enzymes, or indirectly, by changes of the membrane fluidity.

The non-specific interaction of drugs with the lipid moiety may also be of importance for the speed of drug association with proteins. In a manner analogous to that pointed out for the "quicker than diffusion controlled" recombination of lac operon repressor with the operator on the DNA-chain (EIGEN, 1974), the diffusion of drugs in the membrane to the specific binding site of, e.g., cytochrome P450, can be described. The multitude of unspecific sites (see Sect.II.B.) allows the drug to walk around the membrane. The effective area for recombination with cytochrome P450, which depends on the dissociation constant of the unspecific complex and on the diffusion coefficient of the drug, may exceed the size of the active drug binding center of cytochrome P450 by several orders of magnitude. Since the association rate with cytochrome P450 is a function of the encounter distance, the speed of the reaction will be greatly increased when the effective area of combination is enlarged.

Despite an extensive literature on drug interaction with microsomal mixed-function oxidase, only a few papers have been published that deal with the interaction of substances with the microsomal lipid.

Knowledge of how compounds might behave in the lipid phase of microsomes, therefore, has to be derived from experiments with model systems (partition systems, lipid monolayers, bilayers, vesicles) or with intact or modified cell membranes.

A wide variety of structurally dissimilar compounds including tranquilizers, antihistamines, steroids, detergents, vasodilators, sedatives, and narcotics are known to have anesthetic activity, and many of them have been investigated with respect to their binding to, and effects on, different lipid phases (SEEMAN, 1972; BUTLER et al., 1973).

B. Binding of Drugs to Model Systems

Using phospholipid-water systems as a model for biologic membranes, the effects of various drugs on these systems have been examined.

Depending on the chemical structure and charge a drug might:

1. Interact with the polar head groups of the lipids
2. Be intercalated among the lipid alkyl chains
3. At different molecular regions, interact in both ways.

As an example of (3), barbiturates penetrate the lipid layer with their nonpolar portion whereas hydrogen bonds are presumed to form between the polar portion and the phosphates of the phospholipids (NOVAK and SWIFT, 1972).

Electrostatic interactions were shown to be the forces that attract the positively charged nitrogen of L-epinephrine and the negatively charged hydrophilic groups of phosphatidyl-serine. Hydrogen bonding between the phenyl hydroxyls of epinephrine and the phospholipids was further assumed to contribute to the overall binding (HAMMES and TALLMANN, 1971).

The antidepressant drug desiprimine strongly influences the transition temperature of dipalmitoyl-lecithin, but not the transition itself. This implies that the drug acts upon the polar head groups of the lipids and is not positioned between the hydrophobic lipid chains (CHAPMAN, 1973).

In the interaction of the antibiotic chlorothricin with membranes, it has been shown that neither hydrogen nor ionic bonding is an important factor. The action of the drug was interpreted in terms of a nonpolar association of the antibiotic with the alkyl chains of the lipids (PACHE and CHAPMAN, 1972).

The binding of drugs may exert different effects on membrane order and stability. At low concentrations, many lipophilic drugs protect membranes from osmotic, mechanical, or acid lysis (SEEMAN, 1972). However, at high concentrations—usually above 10^{-4} to 10^{-3} M—they are directly or indirectly lytic to the membranes.

The capacity of membranes for drugs is generally high. Under conditions where all sites for the binding of anesthetics are saturated, the membrane concentration of chloropromazine corresponds to one drug molecule per 5 phospholipid molecules (SEEMAN, 1972). An investigation in the author's laboratory (SCHUSTER et al., 1975) with a substituted pleuromutilin showed still higher numbers of about 1 drug per 3 lipid molecules in vesicles from egg yolk lecithin. However, this capacity is strongly influenced by the lipid composition: a predominance of saturated phospholipids in model systems, as well as the presence of cholesterol, reduces the number of sites significantly

without affecting the binding affinity. This dependence on cholesterol is obviously true, also in biologic membranes: Erythrocyte
ghosts and mycoplasma cell membranes with high cholesterol content
exhibit far fewer binding sites than microsomal membranes, which are
poor in cholesterol (SCHUSTER et al., 1975 and unpublished results).

The binding of drugs is accompanied by changes of the membrane properties. In the presence of anesthetics, membranes swell to about 10
times the volume occupied by the drug itself (SEEMAN, 1972). Both
neutral and charged molecules cause a fluidization of the membrane.
For a biologic membrane, the consequences of swelling will range from
a changed permeability to water and other molecules to conformational
changes of membrane enzymes enforced by the drug-induced expansion of
the lipid part.

C. Drug Binding to Microsomal Lipids

ELING and DI AUGUSTINE (1971) observed that binding of the fluorescent
probe 1-anilino naphtalene-8-sulphonate (ANS) to microsomes is largely
attributable to an interaction with membrane phospholipids. They
proposed that the molecule acts in two ways: as an anionic compound,
it interacts electrostatically with the quaternary ammonium of lecithin.
The hydrophobic naphthalene part of the drug will interact with the
nonpolar hydrocarbon chains of the lipids. The fluorescence of ANS-
microsome mixtures was changed in the presence of drugs, being greater
when the drug was in a protonated form (HAWKINS and FREEDMAN, 1973).

In an attempt to describe the binding of the antidepressant drug
imipramine in tissues, GILLETTE (1973) reported that its binding to
liver microsomes could almost entirely be attributed to interactions
with the microsomal phospholipids.

In our laboratory, an extensive study of the microsomal binding of a
substituted pleuromutilin has been undertaken (SCHUSTER et al., 1975).
This antibiotic, consisting of a glycolic acid derivative of a tricyclic diterpene, has been shown to be highly enriched in liver, owing
to microsomal binding. The high capacity of the lipid fraction (about
1 molecule drug per 5 molecules of phospholipid at pH 7.4) together
with an intermediate affinity for binding (K_A about 4 mM^{-1}) accounted
almost completely for the accumulation of the drug in microsomes.
This interaction was further increased at higher pH values when the
drug was in unprotonated form (SCHUSTER, to be published).

Probably, the binding to microsomal lipids influenced the specific
binding to and the metabolism of the substituted pleuromutilin by
cytochrome P450: The type I complex formed at low drug concentration
disappeared, and a new spectral type emerged in parallel to the accumulation of the drug in the lipid phase (see Sect.IV.B.*3*.). This
change was accompanied by a variation in the kinetics of drug hydroxylation. At low concentrations, high specificity of binding was
coupled with a small V_{max}, whereas at high drug concentration, K_m was
identical to the dissociation constant for the drug-lipid complex, and
the maximal velocity was highly increased (SCHUSTER, to be published).

The catalytic activities of a number of further microsomal enzymes like glucose-6-phosphatase and esterases were decreased in parallel to the enrichment of the membranes with the drug (SCHUSTER, to be published).

IV. Interactions of Drugs with Microsomal Proteins

A. General

The contribution of microsomal proteins to the total protein content of an organism is not negligible. Compared with plasma proteins, whose importance in drug binding is well established (GILLETTE, 1973), there is in the human liver about one-third as much microsomal proteins. This multitude of microsomal proteins adhering to the membranes or intimately associated with them provides a great number of potential sites for diverse drug interactions.

As mentioned before, the general properties of only a small number of these proteins have been investigated.

We can assume that drug binding to most proteins will occur in a more or less specific way: since a small segment of plasma proteins is associated with the membranes (cf. Sect.II.B.*3.b.*), microsomes will to a certain degree reflect the interactions of drugs in the plasma.

Of primary importance are those interactions that lead to fewer lipid-soluble structural modifications of a drug. Investigations of those enzyme systems that are responsible for various metabolic alterations of drugs have enormously expanded within the last fifteen years.

Of these proteins particular interest is devoted to the mixed function oxidases. These abundant enzyme systems, with their lack of specificity toward a wide variety of drugs, account for a large number of relatively firm interactions.

Of further importance are enzymes such as esterases or glucuronyl-transferase, which—with an appropriate substrate—may contribute considerably to microsomal protein binding.

A summary of possible drug-protein interactions in microsomes is given in Fig. 1.

B. Interactions of Drugs with Cytochrome P450

1. *Molecular Properties of Cytochrome P450*

Cytochrome P450 is an abundant hemoprotein of the *b*-type, which occurs both in microsomes and mitochondria from various animal tissues (for review see WICKRAMASINGHE, 1975) and in the microsomal fraction of higher plants (RICH, 1975), in bacteria (KATAGIRI et al., 1968),

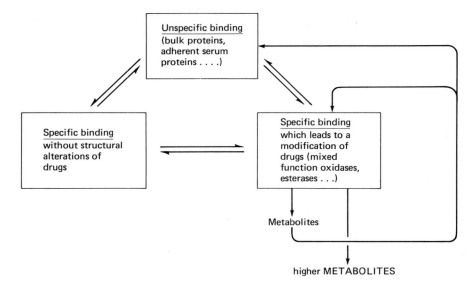

Fig. 1. Interactions of drugs with microsomal proteins

bacteroids (APPLEBY, 1967), and yeast (LEBEAULT et al., 1971). Its
content in liver is far greater than that of all other cytochromes;
in induced animals it accounts for 2-3% of total liver protein (cf.
Sect.II.B.*3.c.* .). Recently, evidence has accumulated that there
exist several multiple forms of the cytochrome that differ in spectral
and catalytic properties (RYAN et al., 1975a). Two of these forms,
cytochrome P450 and P448, which were isolated from rat liver microsomes and purified to a gel electrophoretically homogeneous state,
showed also different subunit weights (RYAN,et al., 1975b). Moreover,
cytochrome P448 itself appears to resemble a class of spectrally
similar hemoproteins, since, in terms of other properties, striking
differences between cytochrome P448 from rabbit and rat liver were
reported (KAWALEK and LU, 1975).

Most of the currently accepted data concerning the properties of cytochrome P450 are derived rrom results obtained with the bacterial cytochrome P450 cam. Since this protein is present in soluble form in
the bacterium *Pseudomonas* cam., it could easily be purified and even
crystallized. The physicochemical properties of this pure enzyme
have been extensively investigated (GUNSALUS et al., 1972; YU et al.,
1974; DUS et al., 1974). Recent attempts to solubilize and purify
the enzyme from hepatic microsomes (RYAN et al., 1975a,b), adrenal
cortex (COOPER et al., 1970), and kidney microsomes (ICHIHARA et al.,
1970) have been successful. A summary of recent data on cytochrome
P450 is given in Table 3.

2. *Structure of the Binding Region*

The term cytochrome P450 is derived from the fact that after reduction
and subsequent addition of carbon monoxide the protein exhibits an
unusual Soret absorption around 450 nm (GARFINKEL, 1958; KLINGENBERG,
1958). All other reduced hemoproteins that bind CO have an absorption

Table 3. Molecular properties of cytochrome P450

Property	Cytochrome P450 cam.	Mammalian cytochrome P450
M (molecular weight in D)	46,000[a] (sedimentation equil.)	280,000[b]
M subunit	46,000[a]	47-49,000 multiple[b] 52-53,000 forms
Prosthethic group	f e r r i p r o t o p o r p h y r i n I X	
Prosthethic group/ subunit	1[a]	1[f]
Binding to apoprotein through polypeptide chains/subunit	acid labile linkage[c] 1[a]	1[f]
Amino acids: Total	397[f]	409[f]
Sum of Pro, Gly, Ala, Val, Met, Ile, Leu, Phe	201[f]	205[f]
SH-groups	6[c]	6[f]
SH-groups titrable	6[d]	
SH-groups involved in heme binding	1[e]	
Contamination	carbohydrate[c]	carbohydrate[f]
Isoelectric point	4.5[c]	4.5[f] (acid carbohydr.)
Immunochemical properties		60-70% cross reactivity[f] against P450 cam antibodies
Substrate	D-camphor and analogues	wide variety of diverse chemical structures
Relative abundance	0.1 nm/mg protein[g]	up to 5 nm/mg microsomal[h] protein (rabbit liver)

[a]YU et al. (1974), [b]VAN DER HOEVEN and COON (1974), [c]DUS et al. (1970), [d]YU and GUNSALUS (1974), [e]TSAI et al. (1970), [f]DUS et al. (1974), [g]PETERSON (1971), [h]ICHIKAWA and YAMANO (1967).

maximum at about 420 nm. This is also the case with denaturated cytochrome P450, the so-called cytochrome P420.

The physical characteristics of hemoproteins depend on the nature of the ligands complexing the central iron. The metal can occur in the ferrous (Fe^{2+}) or the ferric (Fe^{3+}) form with the ligands arranged in a hexavalent octahedral complex. Four positions are occupied by nitrogens of the planar tetrapyrrole. The remaining two ligands above and below the plane are thought to be cysteine (TSAI et al., 1970) and a nonmercaptide group that consists apparently of different ligands endogenous to the protein (STERN et al., 1973). STERN suggested that the mercaptide ligand is more strongly bound than the nonmercaptide ligand. The ferric ion possesses 5 unpaired d-electrons, the ferrous form 6. In a strong ligand field, electrons are forced into orbitals of lower energy, and a low spin state results ($s = 1/2$ for Fe^{3+}, zero for Fe^{2+}). In a weak ligand field, the electrons are distributed over the available orbitals, and a high spin state results ($s = 5/2$ for Fe^{3+}, 2 for Fe^{2+}). Both states are energetically not far apart, and transitions between them can be induced by various factors (BOYD, 1972). The different states can be distinguished by their spectral and magnetic properties. The spectral characteristics of Fe^{3+} cytochrome P450 cam. are given in Table 4.

Table 4. Absorption maxima of Fe^{3+} cytochrome P450 cam. (GUNSALUS et al., 1972)

	Absorption maxima (nm)			
	Soret Band			
High spin	391	500	550	645
Low spin	417	540	570	

Cytochrome P450 from mammalian sources is always isolated in its ferric form (SCHLEYER et al., 1973), which consists of a mixture of both spin states. However, most of cytochrome P450—approximately 95%—was found to occur in the low spin form (STERN et al., 1973). A similar value was also reported for bacterial cytochrome P450 cam. (TSAI et al., 1970).

3. *Binding of Drugs to Cytochrome P450*

The interaction of drugs with microsomal ferric cytochrome P450 causes different types of spectral changes, which are summarized in Table 5.

Table 5. Changes of cytochrome P450-absorption spectra after addition of drugs

	Spectral changes		
	Type I	Type II	Modified Type II
Difference spectrum:			
λ max (nm)	about 390[a] 645[g]	425-435[a]	about 420[f]
λ min (nm)	about 420[a]	390-405[a]	about 390[f]
Absolute spectrum:			
λ max	391[d]	426[e]	
Caused by:	wide variety of compounds that are substrates of the enzyme (drugs, steroids, fatty acids)	basic amines	wide variety of compounds
Interaction:	modification of sixth ligand of heme[b]	directly with heme-iron[a]	
Spin state of complex	high[d]	low[d]	low?[c]

[a]SCHENKMAN et al. (1967), [b]SCHENKMAN and SATO (1968), [c]SCHUSTER et al. (1975), [d]BOYD (1972), [e]REMMER et al. (1969), [f]ORRENIUS et al. (1972), [g]WATERMAN et al. (1973).

a. Type I Binding

Compounds that elicit type I spectra are generally substrates of the drug oxidizing system. Most of the present knowledge on this type of interaction, caused by a variety of structurally different compounds, is derived from experiments with bacterial cytochrome P450 cam. The spectral properties of cytochrome P450 cam.- camphor complex resemble those of mammalian type I complexes (GILLETTE et al., 1972).

Isolated, the enzyme occurs mainly in the low spin state. On addition of D-camphor, the high spin signal increases with increasing substrate up to a value representing 60% of total heme (TSAI et al., 1970). However, titration with camphor indicates a binding of 1 mole substrate per heme (SHARROCK et al., 1973). It was suggested that this spin mixture is caused by an equilibrium of the protein-substrate complex between two different spin states (SHARROCK et al., 1973). The equilibrium between the two states is also valid in the absence of substrate.

The binding of substrate is accompanied by a dramatic change of the absorption spectrum. The Soret maximum is shifted to a lower wavelength, the absorption maxima at 540 and 570 nm are lost, and an additional band at 645 nm, which is thought to be diagnostic for high spin compounds, is observed in bacterial cytochrome P450-camphor complexes as well as in microsomal cytochrome P450-type I complexes (WATERMAN et al., 1973).

The substrate is thought to be located close to the heme iron, since molecular oxygen that is bound to reduced iron must react with the substrate in the enzymatic reaction (PEISACH et al., 1972). PEISACH et al. suggest from EPR-measurements that the substrate is bound in such a way that a direct or indirect asymmetric attachment of an electron-rich part with the heme occurs.

Recently, the kinetics of the relatively firm camphor-binding has been investigated by absorbance measurements with the aid of stopped-flow techniques (GRIFFIN and PETERSON, 1972). The results demonstrate a rapid binding process that follows first order kinetics with respect to both camphor and cytochrome P450. There was no additional reaction observed reflecting a change of protein conformation in the vicinity of heme iron, as is postulated from the conversion of the spin states.

b. Type II Spectra

Type II spectral changes caused by basic amine resemble the formation of a ferri hemochrome, a complex between amine and heme iron (SCHENKMAN et al., 1967); aniline, a typical type II compound, was found to displace carbon monoxide from cytochrome P450. Its interaction with the heme iron is enhanced at acid pH, suggesting the involvement of ionized groups (ionized aniline or protein ligand) in the reaction (ESTABROOK et al., 1972). The binding is competitive with carbon monoxide and presumably with oxygen too (SCHENKMAN et al., 1967). The

complexes are of low spin state, and their spectral properties depend
on the nature of the amines, the maxima in difference spectra varying
from 425 to 435 nm. The ability of substances to elicit type II
spectra is no indication of their hydroxylation by cytochrome P450
(JEFCOATE et al., 1969). Substrates of the monoxygenase also seem to
interact with the type I binding site. In the case of aniline, a
weak type I spectrum is shown to be overlapped by the type II change
(SCHENKMAN, 1970).

c. *Modified Type II Spectra*

This type of spectral change, showing a peak at 420 nm and a trough
at about 390 nm in the difference spectrum, is the mirror image of a
type I spectrum. A large variety of chemically different structures
including substrates of microsomal monoxygenase elicit this spectral
type.

The resulting spectral change is completely different from the type II
change, since there is no competition with aniline and with carbon
monoxide (SCHENKMAN et al., 1972).

Phenacetin and the alkaloid agroclavine are both substrates of the
monoxygenase system, which cause modified type II spectra. However,
at very low drug concentrations type I spectra are observed (ORRENIUS
et al., 1972).

In our laboratory (SCHUSTER et al., 1975), investigations with a substituted pleuromutilin showed that this drug at low concentrations
($<2 \times 10^{-5}$ M) causes type I changes that diminish at increasing concentrations owing to the formation of modified type II interaction.
At high drug concentrations only the latter spectral type is seen.
The disappearance of type I and the formation of modified type II are
paralleled by the unspecific binding of the drug to the phospholipid
part. The modified type II change is not caused by the release of
endogenous type I substrate as was proposed by DIEHL et al. (1970).
Charcoal-treated microsomes were practically devoid of endogenous
substrates. However, they showed type I and modified type II spectra
as did untreated microsomes (SCHUSTER et al., 1977a). Together with
an increasing modified type II-change, a decrease of the high spin
signal at 645 nm occurs (SCHUSTER et al., 1977b).

ORRENIUS et al. (1972) offered 3 alternative explanations for the
existence of modified type II change:

1. It might be due to a displacement of endogenous type I substrate.
2. It might be a composite of type I and type II spectra.
3. It might represent a qualitatively different type of interaction.

From our own results we favor the third alternative. Binding of a
drug to the phospholipid part (to the phospholipids attached to cytochrome P450), might change the membrane structure (the nearest surrounding of cytochrome P450) in such a way that type I interaction
can no longer arise and a different type of interaction occurs.

4. *Hydroxylation of Drugs*

a. *Components of Mixed Function Oxidase*

The cytochrome P450-containing electron transport chains from diverse
sources differ with regard to constituents and their structural
organization.

In mammalian microsomes, the essential components of the chains are
tightly bound to the membranes in a nonstoichiometric relationship,
whereas in bacterial systems all components occur in stoichiometric
proportions and in soluble form (ESTABROOK et al., 1972).

The enzyme system solubilized from liver microsomal membranes has been
resolved into three functional components (LU and COON, 1968):

1. The key enzyme cytochrome P450 binds the drug as previously discussed and transfers molecular oxygen to it.
2. NADPH-dependent cytochrome P450 reductase transfers reducing equiv-
alents from NADPH to the cytochrome (ESTABROOK et al., 1972).
3. A heat stable lipid fraction, identified as phosphatidyl choline,
seems to be essential for the electron transfer (STROBEL et al.,
1970).

The coupling between reductase and cytochrome P450 proves to be not
highly specific, the lipid fraction conferring no specificity to the
system. In a reconstituted system capable of drug hydroxylation, the
reductase can be replaced by artificial photochemical systems or by
the xanthine/xanthine oxidase complex (COON et al., 1973).

Cytochrome P450 isolated from such varied sources as rabbit liver (LU
et al., 1969b), human liver (KASCHNITZ and COON, 1972), and yeast
(DUPPEL et al., 1973) has been effectively coupled with the reductase
from rat liver microsomes and phospholipid from the same source or
with a synthetic phospholipid.

Cytochrome P450 preponderates over the reductase. It is assumed that
one reductase molecule reacts with about 12 cytochrome molecules
(ESTABROOK et al., 1972).

Recently, ROGERS and STRITTMATTER (1974) proposed a model for the
localization of the cytochrome b5-cytochrome-b5-reductase system in
microsomal membranes. They suggest that both enzymes are randomly
distributed on the membrane and undergo lateral diffusion within the
membrane. A similar structural organization for cytochrome P450 and
its reductase was proposed by YANG (1975). He assumes that a non-
rigid association of the enzymes, together with lateral mobility,
will enable the reductase to react efficiently with all cytochrome
P450 molecules.

b. *Participation of Cytochrome b5*

Several authors (COHEN and ESTABROOK, 1971; CORREIA and MANNERING,
1973; HILDEBRANDT and ESTABROOK, 1971) suggest participation of the

second microsomal electron transport system cytochrome b5/NADH-dependent cytochrome b5 reductase in drug hydroxylation. It has been shown that NADH in the presence of NADPH has a synergistic effect on the metabolism of a variety of substrates. The interactions between both electron transport chains are thought to consist of a transfer of the second reduction equivalent to cytochrome P450 via cytochrome b5.

Investigations with the purified reconstituted cytochrome P450 complex in the absence of any detectable cytochrome b5 indicate that cytochrome b5 is not an indispensable component of the hydroxylation system (LU et al., 1974; LEVIN et al., 1974). Moreover, antibodies against cytochrome b5 do not inhibit the reduction of cytochrome P450 in microsomes or the metabolism of aniline or pyramidon (SASAME et al., 1973), whereas they decrease significantly the cytochrome b5-dependent reaction as the desaturation of fatty acids (OSHINO and OMURA, 1973).

The role of the cytochrome b5 system in drug hydroxylation may be confined to two interactions with the cytochrome P450 system (LU et al., 1974), namely, to a regulation of the NADPH-dependent reactions and to the involvement in a slow, but measurable NADH-dependent hydroxylation.

c. Hydroxylation Mechanism

A wide variety of compounds both endogenous and foreign serve as substrates for cytochrome P450. The metabolism of these molecules can apparently be regarded as a series of diverse oxidative reactions. A collection of these reactions, which resemble different kinds of hydroxylation, together with a few selected substrates, is presented in Table 6.

The overall reaction involving molecular oxygen was formulated as

$$RH + O_2 + XH_2 \rightarrow ROH + H_2O + X \text{ (HAYAISHI, 1964)}$$

where RH represents substrate and XH_2, the electron donor.

Table 6. Hydroxylation reactions

Reaction	Substrate	References
Aliphatic hydroxylation	cyclohexane	ULLRICH, 1969
	n-butane, isobutane	
	n-pentane	FROMMER et al., 1970
Aromatic hydroxylation	aniline	BAUER and KIESE, 1964
	3,4 benzpyrene	CONNEY et al., 1957
N-Dealkylation	ethylmorphine	DAVIES et al., 1969
	codein	AXELROD, 1956a
	aminopyrine	REMMER et al., 1966
O-Dealkylation	chinine, codein, cholicin	AXELROD, 1956b
Deamination	amphetamine	AXELROD, 1955
Sulfoxidation	chlorpromazine	GILLETTE and KAMM, 1960
N-Oxidation	aniline, sulphanilamide,	
	phenetidine	UEHLEKE, 1971

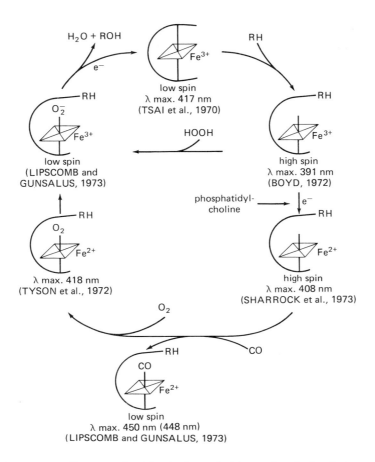

$H_2O + ROH$

e^-

Fe^{3+}

low spin
λ max. 417 nm
(TSAI et al., 1970)

RH

RH

RH

O_2^-

Fe^{3+}

low spin
(LIPSCOMB and
GUNSALUS, 1973)

HOOH

Fe^{3+}

high spin
λ max. 391 nm
(BOYD, 1972)

RH

O_2

Fe^{2+}

λ max. 418 nm
(TYSON et al., 1972)

phosphatidyl-
choline

e^-

RH

Fe^{2+}

high spin
λ max. 408 nm
(SHARROCK et al., 1973)

O_2

RH

CO

CO

Fe^{2+}

low spin
λ max. 450 nm (448 nm)
(LIPSCOMB and GUNSALUS, 1973)

Fig. 2. Sequence of steps in the hydroxylation of substrates

Studies of the bacterial camphor hydroxylating enzyme system together
with increasing information on the purified reconstituted microsomal
cytochrome P450 complex have contributed largely to present knowledge
of the mechanism of hydroxylation. An essential step in that pathway
is the cleavage of the oxygen molecule in such a way that one oxygen
atom is transferred to the substrate and the other reduced to water,
and the transfer of two electrons. It is unknown whether electron
transfer precedes or follows the step in which oxygen reacts with the
substrate (COON et al., 1973). The reaction mechanism as proposed by
COON is depicted in Fig. 2.

The binding of the substrate to the enzyme is rapid, as previously
discussed, changing the state of the heme iron to high spin. The
subsequent reduction of the binary complex is mediated by NADPH-
dependent cytochrome P450 reductase (LU et al., 1969a). As could be
shown by stopped flow kinetics, the presence of phosphatidylcholine
greatly increases the reduction rate (STROBEL et al., 1970), although
the manner in which the lipid facilitates the electron transfer is
unknown to date.

In the absence of the substrate, the reduction of cytochrome P450 is remarkably slower (DIEHL et al., 1970). The correlation of cytochrome P450 reduction with the overall rate of drug metabolism makes it probable that this step is rate-limiting in the whole scheme (GIGON et al., 1969). The binding of oxygen to the binary complex is rapid and leads to a ternary complex with absorption maximum at 418 nm (TYSON et al., 1972). Rapid scan stopped flow studies show that this oxy P450-form predominates during the hydroxylation reaction.

Reduced cytochrome P450 reacts alternatively and also rapidly with carbon monoxide. This complex, with its absorbance maximum at 450 mμ, is frequently used as an indicator for the reduction reaction. Evidence for the formation of a ternary complex between reduced cytochrome, substrate and carbon monoxide has been obtained by equilibrium dialysis of enzyme and ^{14}C-labeled camphor in the presence of excess CO. The binding of one camphor molecule per cytochrome without significant change in the affinity suggests that the binding of CO does not interfere significantly with substrate binding (LIPSCOMB and GUNSALUS, 1973).

It is suggested that the ternary complex of cytochrome P450 with substrate and oxygen accepts a second electron required for the reduction of a single atom of oxygen, while the other oxygen atom is transferred to the substrate. This series of steps might well be concerted reactions that remain as yet uncharacterized. COON et al. (1973) proposed two alternatives: that molecular oxygen undergoes enzymatic activation (via reduction) prior to reaction with the substrate and that molecular oxygen undergoes an enzyme-catalyzed reaction with the substrate prior to reduction. There is evidence that a peroxidative mechanism operates. Hydrogen peroxide is generated during NADPH-oxidation (HILDEBRANDT, 1973). It was shown by COON et al. (1975) that highly purified cytochrome P450 from mammalian liver catalyzes drug hydroxylation in the presence of hydrogen peroxide and the absence of NADPH and NADPH-dependent reductase (shown as an abbreviated cycle in Fig. 2).

The hydroxylated, less lipophilic product shows a changed binding to the cytochrome. Hydroxylated camphor, e.g., does not shift the position of the Soret band any more; however, it exhibits a binding constant similar to that of camphor (GUNSALUS, 1973). Products are either released from the enzyme or they can serve as substrates for further hydroxylation processes.

C. Involvement of Cytochrome b5 in Microsomal Drug Interactions

Cytochrome b5 obviously participates in microsomal processes solely as an intermediary electron carrier (OSHINO et al., 1971). A scheme for the part it plays in fatty acid desaturation was given by HOLLOWAY and KATZ (1972) and is shown in Fig. 3.

Cytochrome b5 consists of a hydrophobic tail, buried within the reticular membrane, and a globular head exposed to the cytoplasma, which contains the heme group in a hydrophobic pocket (MATHEWS et al., 1971).

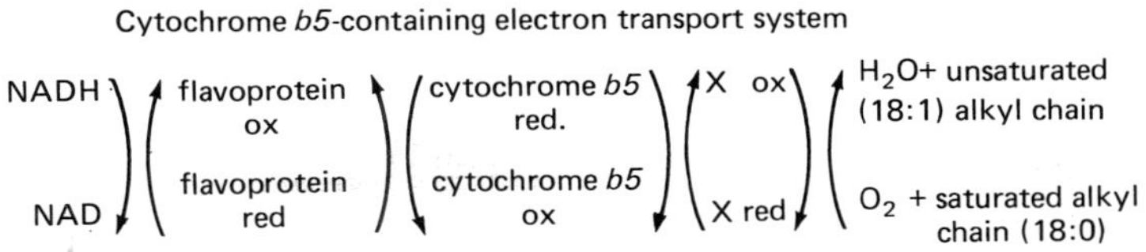

Cytochrome *b5*-containing electron transport system

$$NADH \rbrace \lbrace \uparrow \begin{matrix} flavoprotein \\ ox \\ flavoprotein \\ red \end{matrix} \rbrace \lbrace \begin{matrix} cytochrome \, b5 \\ red. \\ cytochrome \, b5 \\ ox \end{matrix} \rbrace \lbrace \uparrow \begin{matrix} X \quad ox \\ X \, red \end{matrix} \rbrace \lbrace \uparrow \begin{matrix} H_2O+ \, unsaturated \\ (18:1) \, alkyl \, chain \\ O_2 + \, saturated \, alkyl \\ chain \, (18:0) \end{matrix}$$

Fig. 3. Cytochrome b5-containing electron transport system

The iron atom of cytochrome b5, unlike that in cytochrome P450, is rather inert and will not bind foreign ligands, while the protein is in its native conformation (MATHEWS et al., 1971). This is due to the fact that the fifth and sixth ligands of iron are histidines that are held firmly in place by a variety of interactions with the main and the side chains of the protein (MATHEWS et al., 1971). Any substitution of histidine will produce a major disruption of the protein structure. For this reason, cytochrome b5 will not be able to act as a terminal oxidase, but will be restricted to interactions with the NADH-dependent reductase (STRITTMATTER and VELICK, 1957) and further with an oxidase indicated by X in Fig. 3.

X-ray analysis of cytochrome b5 (MATHEWS et al., 1971) up to 2.0 Å resolution showed a hydrophobic groove on the surface of the protein Most of the hydrophobic residues in the groove can be removed by tryptic cleavage. The remaining peptide still binds the heme group, but is no longer able to interact with the reductase. MATHEWS et al. (1971) concluded that this hydrophobic groove might be the site of interaction with the reductase.

SATO et al. (1969) defined the protein as a relay station where electrons from NADH, NADPH, and ascorbate are passed to a cyanide-sensitive factor (X in Fig. 3), which activates oxygen for the desaturation process.

Besides a possible involvement of cytochrome b5 in drug hydroxylation mediated by cytochrome P450, where the second electron is assumed to come from cytochrome b5 (see Sect.IV.B.*4.b.*), the desaturation of alkyl chains seems to be another prominent reaction of cytochrome b5.

Recently, participation of cytochrome b5 and its reductase was detected in the reduction of N-hydroxylamines (KADLUBAR and ZIEGLER, 1974). However, a third protein component of unknown function is required for activity. Obviously, this protein does not contain an electron transfer group undergoing oxidation and reduction.

D. Interactions of Drugs with the Mixed Function Amine Oxidase

ZIEGLER and PETTIT (1966) described an enzyme catalyzing N-oxidation of dimethylaniline that required NADPH and molecular oxygen and was insensitive to carbon monoxide and the potent type I binding inhibitor SKF 525 A. From this fact, they concluded that cytochrome P450 was not an obligatory component in the reaction sequence.

Dimethylaniline N-oxidase activity is concentrated in the microsomal fraction of liver. ZIEGLER and MITCHELL (1972) succeeded in isolating the enzyme from pig liver microsomes and purifying it to a high degree.

The isolated oxidase was shown to be composed of similar, if not identical subunits, containing FAD (ZIEGLER and MITCHELL, 1972), and to be immunologically distinct from NADPH cytochrome c-reductase (MASTERS and ZIEGLER, 1971). Its main substrates are secondary and tertiary amines, although a few primary aromatic amines (e.g., 1-naphthylamine and 2-naphthylamine) are also metabolized by the enzyme (BECKETT, 1971; ZIEGLER et al., 1971). Tertiary amines are oxidized to N-oxides, secondary amines, to hydroxylamines.

From kinetic experiments, ZIEGLER et al. (1971) concluded that the enzyme contains at least two sites that interact with amines: a catalytic site and a regulatory site. Many secondary and tertiary alkyl amines react with both sites, but primary alkylamines were found to activate the enzyme by interacting with the regulatory site.

In addition to primary amines, a number of other types of compounds, especially those containing a guanidine group, also showed activating properties (ZIEGLER et al., 1973).

Recently, it has been found that the enzyme system is also capable of S-oxidation (POULSEN et al., 1974). In addition to the oxidation of thioureylenes, the enzyme catalyzes also the oxidation of a few sulfhydryl compounds like dithiothreitol and its cyclic disulfide.

E. Interaction of Drugs with Esterases

A considerable number of enzymes are known that exhibit esterase activity and appear to be ubiquitous in the cell. However, the esterase activities of, e.g., L-chymotrypsin, 3-phosphoglyceraldehyde dehydrogenase, and aldehyde dehydrogenase are several orders of magnitude lower than those of real esterases (JUNGE and KRISCH, 1973). Liver esterases comprise a group of different multiple forms that are associated primarily in a loose attachment to the microsomal fraction (KRISCH, 1971). The enzymes are able to hydrolyze ester linkages of carboxylic- and thioesters as well as amide bonds. A selection of compounds that serve as substrates is given in Table 7.

Esterases also catalyze the transfer of acyl groups. Ethyl acetate and methyl butyrate have been shown to be suitable acyl donors for the acylation of aniline and p-phenetidine (FRANZ and KRISCH, 1968b). The reaction of the enzymes with certain amino acid esters leads finally to the formation of dipeptides (BENÖHR and KRISCH, 1967).

Among carboxylesterases, the enzyme derived from pig liver is the best defined. The highly purified protein was shown to give at least 4 bands in gel electrophoresis, which exhibited differences toward various esterase substrates (JUNGE and KRISCH, 1973). However, no separation of amidase and carboxylic esterase was achieved, thus providing evidence for the identity of unspecific carboxylesterase with acetanilide amidase.

Table 7. Linkages that are hydrolyzed by microsomal esterases

Linkage	Compound	References
Carboxylic ester	tributyrin, *p*-nitro-phenylacetate	FRANZ and KRISCH, 1968
	tyrosine ethylester, procaine, *n*-butyl-*n*-butyrate	ARNDT and KRISCH, 1973
Thioester	*p*-nitrothiophenyl-hippurate	STOOPS et al., 1969
	thiophenyl acetate	GREENZAID and JENCKS, 1971
Aromatic amides	acetanilide, phenacetin	FRANZ and KRISCH, 1968a
	L-leucyl-*β*-naphthyl-amide	BERNHAMMER and KRISCH, 1966
	hostacaine	ARNDT and KRISCH, 1973

Pig liver carboxylesterase was shown to be a trimer bearing one active site per subunit (KUNERT and HEYMANN, 1975). The binding site for substrates is obviously a lipophilic pocket that is either large or flexible enough to bind higher alkyl groups of different chain lengths (ARNDT and KRISCH, 1973). The mechanism of ester hydrolysis is assumed to proceed via an acyl-enzyme intermediate. Organophosphorus compounds, such as diethyl-*p*-nitrophenyl phosphate, form analogously a complex with the enzyme, which, however, cannot be hydrolyzed. These compounds have been used for affinity labeling of the active site, and the amino acid sequence around an active serine, where the inhibitor is bound, was determined (KRISCH, 1971).

Besides the binding of substrates to the active site, ARNDT and KRISCH (1973) propose further unspecific sites on the protein that can either be occupied by excess substrate molecules or by other lipophilic agents. The interaction with these sites was suggested to lead to alterations of groups essential for enzymatic activity.

F. Interactions of Drugs with Glucuronyl Transferase

Since glucose is generally available in biologic systems, the formation of glucuronides is one of the more common routes in terminal drug metabolism. Owing to its large size and high glucuronyl transferase activity, the liver appears to be the main site for glucuronidation (MIETTINEN and LESKINEN, 1970). The conjugation reaction is carried out by microsomal glucuronyl transferase, which binds UDP-glucuronic acid and transfers it to suitable acceptors, usually hydroxyl-, carboxyl-, amino-, and sulfhydryl-groups (DUTTON, 1971). Glucuronidation is frequently preceded by microsomal hydroxylation reactions.

A summary of substrates for glucuronidation is given in Table 8.

Microsomal glucuronyl-transferase appears to consist of a group of closely related enzymes or at least to possess separate aglycone sites

Table 8. Substrates for glucuronyl transferase (WILLIAMS, 1967; LAYNE, 1970)

Type of acceptor group	Substrate
Hydroxyl-	*tert*. butyl alcohol, phenol 4-hydroxycoumarin N-hydroxy-2 acetylamino fluorene testosterone, estrone, estradiol
Carboxyl-	benzoic acid phenylacetic acid
Amino-	aniline sulfadimethoxine sulfisoxazole
Sulfhydryl-	2-mercaptobenz-thiazole N,N-diethyldithio carbamic acid

for the conjugation of *p*-nitrophenol, *o*-aminophenol, and *o*-aminobenzoic acid (ZAKIM et al., 1973a).

The enzyme is intimately associated with the microsomal membrane and can be solubilized by trypsin digestion and digitonin treatment (PUUKKA et al., 1975). The phospholipid environment in the natural membrane exerts a constraint on the enzymatic activity (ZAKIM et al., 1973b). The relief of this constraint, e.g., by treatment of microsomes with phospholipase, is associated with a loss of specificity in the binding of UDP-glucuronic acid. Since the concentration of UDP-glucuronic acid in liver is low compared to the concentration required *in vitro* for half maximal enzymatic activity (ZAKIM et al., 1973b), an activation of the enzyme could adapt it to higher turnovers in the presence of excess aglycones. ZAKIM et al. (1973b) proposed that the phospholipid moiety may take part in the regulation of the conjugating activity.

V. Effects of Drugs on Microsomal Constitution and Activities

A. Induction of Microsomal Enzymes

The induction of microsomal enzymes leads to an accelerated metabolism of drugs and other foreign and endogenous lipophilic compounds, and hence, changes the intensity and the duration of drug action in the organism. The phenomenon of induction was first described by BROWN et al. (1954), who investigated dietary influences on the demethylation of amino-azo dyes. In the subsequent years, a large number of compounds including drugs, carcinogens, insecticides, and steroids were found to stimulate the activity of drug-metabolizing enzymes (Table 9). Most of these substances are of lipophilic character (for review see CONNEY, 1967; REMMER, 1969).

Table 9. Inducers of microsomal activities (REMMER, 1969)

Pharmacologic action	Compound acting as inducer
Hypnotics and sedatives	urethane, barbiturates, glutethimide, pyridione, chlorbutanol
Anticonvulsants	diphenylhydantoin, paramethadione, primidone
Tranquillizers	chlorpromazine, triflupromazine, .chlordiazepoxid
Analgesics	phenylbutazone, aminopyrine
Muscle relaxants	meprobamate, phenaglycodol
Antihistaminics	orphenadrine, diphenhydramine, chlorcyclizine
Central nervous system stimulants	nikethamide, bemegride
Psychomotor stimulants	imipramine, iproniazid
Hypoglycemic agents	tolbutamide, chlorpropamide, carbutamide
Steroids	nortestosterone, norethandrolone, norethynodrol
Insecticides	chlordane, DDT, hexachlorcyclohexane dieldrin, aldrine, heptachlor
Carcinogens	3-methylcholanthrene, 3,4-benzpyrene

At least two types of inducers can be distinguished: (1) those belonging to the phenobarbital class, and (2) those belonging to the 3-methylcholanthrene class. Both groups stimulate various pathways of drug metabolism, such as oxidation reduction reactions, glucuronide formation, and de-esterification. However, compounds of the 3-methylcholanthrene type apparently induce a more limited group of reactions (CONNEY, 1967). A comparison of the effects exerted on microsomes by both inducing groups is given in Table 10.

Induction phenomena are restricted to intact cell systems. Addition of inducers to liver homogenate does not increase the activity of metabolizing systems (CONNEY et al., 1957, 1960).

The increase in the amount of microsomal enzyme components can be caused by an increased rate of protein synthesis and/or decreased rates of breakdown. Inhibitors that block protein synthesis by different mechanisms, e.g., ethionine (CONNEY et al., 1960), puromycin and actinomycin D (ORRENIUS et al., 1965), have been shown to abolish the inducing effect of drugs. Turnover studies carried out by GREIM (1970) demonstrated that the increased amounts of cytochrome P450 are a result of both an enhanced synthesis of the protein and an inhibition of its breakdown.

The complicated sequence of events including enhanced synthesis of proteins and phospholipids, which leads to proliferation of the reticulum and an enlargement of the whole organ, is known to date in some detail, but still too incompletely for a molecular basis of induction

Table 10. Effects of phenobarbital and 3-methylcholanthrene treatment on liver microsomes

Effect	After treatment with	
	phenobarbital	3-methylcholanthrene
Increase in cytochrome P450	+[a]	+[a] (various form: cytochrome P448)
cytochrome P450 reductase	+[a]	-[a]
cytochrome P450 reduction rate	+[a]	-[a]
Increase of total liver protein	+[c,d]	+[c]
protein per unit weight of liver tissue	+[c,d]	-[c]
Proliferation of smooth reticulum	marked[e,f]	small[f]
Change of lipid protein interactions (indicated by ANS-fluorescence and solubilization methods)	+[b]	-[b]

[a]GILLETTE (1971), [b]LAITINEN et al. (1974), [c]CONNEY and GILMAN (1963), [d]CONNEY et al. (1960), [e]REMMER and MERKER (1963), [f]FOUTS and ROGERS (1965).

Table 11. Sequence of events after application of an inducer (phenobarbital class)

Time after injection (h)	Event
0-3	drug binding[a]
3	increased rate of phospholipid turnover[b]
6	increase in hydroxylating enzyme system in rough reticulum[c]
8-12	increase in phospholipid content, increase in RNA-polymerase content[d]
until 24	increase of enzyme content in smooth reticulum, proliferation of smooth membranes and of total liver[d]

[a]ERNSTER and ORRENIUS (1965), [b]ORRENIUS et al. (1965), [c]ORRENIUS (1965), [d]ORRENIUS et al. (1969).

to be formulated. The sequence of events after injection of an inducer is summarized in Table 11.

The stimulating effect of drugs on their own metabolism has various pharmacologically important consequences:

1. During prolonged treatment drugs become less effective. Compounds are metabolized more quickly, and hence, attain lower plasma levels.
2. If a metabolite has a higher therapeutic activity than the parent compound, induction can lead to enhanced drug action.
3. If toxic metabolites or very reactive metabolites are formed, induction can cause increased toxicity of the drug. The induction caused by a single drug will lead to increased metabolism and a modified action of endogenous substrates like steroids.

B. Toxicity of Drug Metabolites

The pathways of microsomal drug metabolism lead in most cases to products that are less toxic and more polar than the parent substances and that can easily be excreted. However, a number of routes may also result in the formation of chemically reactive metabolites that are able to combine covalently with tissue molecules. A few examples of the formation of chemically reactive metabolites are given in Table 12.

Target substances for reactive metabolites were found to be RNA, DNA, proteins (for review see: MILLER and MILLER, 1966), and lipids (REYNOLDS, 1967). In fact, the very small amounts of drug that are found as covalently bound residue have made it difficult to investigate the nature of binding sites for many drugs. It is also unknown

Table 12. Formation of chemically reactive metabolites (DALY et al., 1972; WEISBURGER and WEISBURGER, 1973; MAGEE and SCHOENTHAL, 1964; MILLER and MILLER, 1966; UEHLEKE, 1969; GILLETTE et al., 1974; SHANK, 1975; OESCH, 1972)

Parent compound	Reactive product

Primary amines
  β-naphthyl amine
  4-aminobiphenyl
  benzidin
  4-aminostilbene
  2-aminoanthracene
Secondary amines
  N-methyl-4-aminobenzene
  2-acetylaminofluorene
Amino-azo-dyes
  N-N-dimethyl-4-amino-
    azobenzene
Carbamate
  urethane
	N-hydroxylated products (further activation, e.g., by formation of N-O sulfate esters)

N-Nitroso compounds
  dimethyl nitrosamine
  methylbenzyl nitrosamine
  methylnitroso urethane
  methyl nitroso urea
  cycasin
	hydroxylated nitrosamine alkyl carbonium ion

Polycyclic hydrocarbons
aromatic
  naphthalene
  dibenzanthracene
  phenanthrene
  pyrene
  benzopyrene
  halobenzenes
olefinic
  dieldrin
  allyl substituted barbiturates
  carbamezepine
  styrene
  cyclohexene
  steroids
	epoxides

to what extent covalent binding has to occur in order to produce serious toxic effects such as carcinogenesis, mutagenesis, necrosis, or fetotoxicities.

For a small number of compounds, a correlation between hepatotoxicity and the extent of covalent binding of metabolites was described recently:

In the case of halobenzenes, which are partially metabolized to epoxides, it has been shown that covalent binding of these products parallels liver necrosis (BRODIE et al., 1971). This binding of bromobenzene-product was strongly decreased by the presence of excess gluthathione. In addition to other pathways—e.g., reaction with microsomal epoxide hydrase (OESCH, 1972)—epoxides can be converted to inert products by conjugation with gluthathione, a reaction that takes place in the soluble fraction of liver cell (BRODIE et al., 1971). However, above a threshold dose of bromobenzene, the levels of gluthathione are depleted, and considerable covalent binding occurs (GILLETTE et al., 1974). Pretreatment of rats with phenobarbital increases the rate of bromobenzene metabolism and also the level of free reactive epoxides. At high doses of bromobenzene, covalent binding is markedly increased together with the severity of liver necrosis (REID and KRISHNA, 1973).

The analgesic acetaminophen was shown to produce severe hepatic necrosis when a high overdose was taken (GILLETTE et al., 1974). The toxic effect is due to the cytochrome P450-mediated formation of a reactive metabolite presumably a N-hydroxylation product. Gluthathione forms a nontoxic conjugate with the metabolite, which is readily excreted as mercapturic acid (MITCHELL et al., 1973). Significant covalent binding of the metabolite to nucleophilic groups in liver proteins occurred only with large doses of acetaminophen, which depleted hepatic gluthathione (MITCHELL et al., 1973).

Finally, a dose-dependent hepatotoxicity caused by the covalent binding of a reactive metabolite was also found for the drug furosemide (GILLETTE et al., 1974). In this case, however, gluthathione did not appear to be protective, since covalent binding occurred also in the presence of high levels of gluthathione.

VI. Summary and Concluding Remarks

The main part of this survey has been devoted to the crucial role of single microsomal enzyme systems in drug metabolism. Their degree of unspecificity together with the fact that drugs and other environmental chemicals can induce most of these enzymes adapts the organism to transform rapidly novel lipophilic compounds into water-soluble structures.

However, a description of drug interactions with microsomes membranes must also take into account that:

1. Due to their lipophilicity, many drugs are dissolved in the lipid bulk and—as has been shown with some biologic systems—alter membrane structure (SEEMAN, 1972).
2. Among other proteins, drug metabolizing enzymes are integral membrane constituents sensitive to lipid fluidity (ELETR et al., 1973).

It has been suggested (TRÄUBLE and EIBL, 1975) that the function of membrane proteins depends on their rotational or lateral motion, which is inhibited when lipid motion is reduced.

Although microsomes are mainly in a fluid state at physiologic temperature, islets of rather rigid structure remain. It was proposed that phospholipid halos of crystalline structure enclose the cytochrome P450-cytochrome P450 reductase system (STIER and SACKMANN, 1973).

The binding of drugs to the lipid moiety can trigger a phase transition (CHAPMAN, 1973) in the lipid islets around membrane proteins. In addition to lateral expansion of these parts of the membrane, rapid lateral diffusion of lipid molecules will occur. The resulting fluid state will directly or indirectly affect the conformation of proteins by enhancing their freedom of motion.

An effect on protein conformation and enzymatic activity can also be exerted by drugs that bind to unspecific sites on microsomal proteins, as was proposed by ARNDT and KRISCH (1973) for esterases.

Finally, the unspecific interactions with a large number of binding sites in the lipid bilayer will allow a drug to "walk" rapidly around the endoplasmic network: (1) providing thus a continuous supply for the transforming enzymes and; (2) possibly enhancing the association rates with these enzymes by enlarging the area of recombination (EIGEN, 1974).

A diagrammatic summary of possible drug interaction with microsomal membranes is shown in Fig. 4.

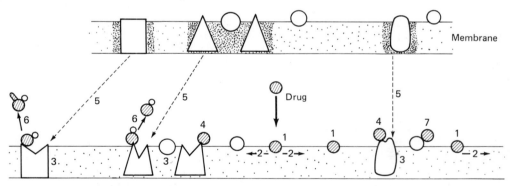

Fig. 4. Drug interactions with microsomal membranes: (1) Drug binding to lipid phase; (2) Diffusion of drug in lipid phase; (3) Fluidization of rigid lipid islets (expansion of membrane); (4) Specific binding of drugs to proteins (effect of lipid binding on association rate); (5) Change of protein conformation; (6) Metabolism of drug, release of product; (7) Unspecific binding of drugs to proteins

References

APPLEBY, C.A.: A soluble haemprotein P450 from nitrogen-fixing rhizobium bacteroids. Biochim. Biophys. Acta 147, 399-402 (1967).
ARNDT, R., KRISCH, K.: Catalytic properties of an unspecific carboxyl-esterase (E1) from rat liver microsomes. Europ. J. Biochem. 36, 129-134 (1973).

AXELROD, J.: The enzymatic deamination of amphetamine (benzedrine). J. Biol. Chem. 214, 753-763 (1955).

AXELROD, J.: The enzymatic N-demethylation of narcotic drugs. J. Pharmacol. Exp. Ther. 117, 322-330 (1956a).

AXELROD, J.: The enzymic cleavage of aromatic ethers. Biochem. J. 63, 634-639 (1956b).

BAUER, S., KIESE, M.: Heterogeneousness of the microsomal enzymes effecting the o- and p-hydroxylation of aniline. Naunyn-Schmiedeberg's Arch. Exp. Pathol. Pharmak. 247, 144-148 (1964).

BECKETT, A.H.: Metabolic oxidation of aliphatic basic nitrogen atoms and their α-carbon atoms. Xenobiotica 1, 365-384 (1971).

BENÖHR, H.C., KRISCH, K.: Carboxylesterase aus Rinderlebermikrosomen. Isolierung, Eigenschaften und Substratspezifität. Z. Physiol. Chem. 348, 1102-1114 (1967).

BERNHAMMER, E., KRISCH, K.: Zur Hydrolyse von Aminosaure-Arylamiden durch mikrosomale Schweineleberesterase und Serum. Z. Klin. Chem. 4, 49-55 (1966).

BIELKA, H.: Endoplasmatisches Retikulum und Ergastoplasma. In: Molekulare Biologie der Zelle (Ed. H. BIELKA), pp. 399-432. Stuttgart: G. Fischer, 1969.

BOYD, G.S.: Biological hydroxylation reactions. In: Biological Hydroxylation Reactions (Eds. G.S. BOYD, R.M.S. SMELLIE), pp. 1-9. London-New York: Academic Press, 1972.

BRODIE, B.B., REID, W.D., CHO, A.K., SIPES, G., KRISHNA, G., GILLETTE, J.R.: Possible mechanism of liver necrosis caused by aromatic organic compounds. Proc. Nat. Acad. Sci. 68, 160-164 (1971).

BROWN, R.R., MILLER, J.A., MILLER, E.C.: The metabolism of methylated aminoazo dyes. IV Dietary factors enhancing demethylation *in vitro*. J. Biol. Chem. 209, 211-222 (1954).

BUTLER, K.W., SCHNEIDER, H., SMITH, I.C.: The effects of local anaesthetics on lipid multilayers. A spin probe study. Arch. Biochem. Biophys. 154, 548-554 (1973).

CAMPBELL, P.N., LAWFORD, G.R.: The protein synthesizing activity of the endoplasmic reticulum in liver. In: Structure and Function of the Endoplasmic Reticulum in Animal Cells (Ed. F.C. GRAN), pp. 57-79. Oslo: Univ. Forlaget, 1968.

CHAPLIN, M.D., MANNERING, G.J.: Role of phospholipids in hepatic microsomal drug-metabolizing system. Mol. Pharmacol. 6, 631-640 (1970).

CHAPMAN, D.: Some recent studies of lipids, lipid-cholesterol and membrane systems. In: Biological Membranes (Ed. D. CHAPMAN, D.F.H. WALLACH), Vol. 2, pp. 91-144. London-New York: Academic Press, 1973.

CLAUDE, A.: Fraction from normal chick embryo similar to tumor producing fraction of chick tumor I. Proc. Soc. Exp. Biol. Med. 39, 398-403 (1938).

CLAUDE, A.: Constitution of protoplasm. Science 97, 451-456 (1943).

CLAUDE, A.: Studies on cells: Morphology, chemical constitution and distribution of biochemical functions. Harvey Lect. 43, 121-164 (1948).

CLAUDE, A.: Microsomes, endoplasmic reticulum and interactions of cytoplasmic membranes. In: Microsomes and Drug Oxidation (Eds. J.R. GILLETTE, A.H. CONNEY, G.J. COSMIDES, R.W. ESTABROOK, J.R. FOUTS, G.J. MANNERING), pp. 3-39. New York-London: Academic Press, 1969.

CLAUDE, A., PORTER, K.R., PICKELS, E.G.: Electron microscope study of chicken tumor cells. Cancer Res. 7, 421-430 (1947).

COHEN, B.S., ESTABROOK, R.W.: Microsomal electron transport reactions. II. The use of reduced triphosphopyridine nucleotide and/or reduced diphosphopyridine nucleotide for the oxidative N-demethylation of aminopyrine and other drug substrates. Arch. Biochem. Biophys. 143, 46-53 (1971).

CONNEY, A.H.: Pharmacological implications of microsomal enzyme induction. Pharmacol. Rev. 19, 317-366 (1967).

CONNEY, A.H., DAVISON, C., GASTEL, R., BURNS, J.J.: Adaptive increases in drug metabolizing enzymes induced by phenobarbital and other drugs. J. Pharmacol. Exp. Ther. 130, 1-8 (1960).

CONNEY, A.H., GILMAN, A.G.: Puromycin inhibition of enzyme induction by 3-methylcholanthrene and phenobarbital. J. Biol. Chem. 238, 3682-3685 (1963).

CONNEY, A.H., MILLER, E.C., MILLER, J.A.: Substrate induced synthesis and other properties of benzpyrene hydroxylase in rat liver. J. Biol. Chem. 228, 753-766 (1957).

COON, M.J., NORDBLOM, G.D., WHITE, R.E., HAUGEN, D.A.: Purified liver microsomal cytochrome P450: Catalytic mechanism and characterization of multiple forms. Biochem. Soc. Trans. 3, 813-817 (1975).

COON, M.J., STROBEL, H.W., BOYER, R.F.: On the mechanism of hydroxylation reactions catalyzed by cytochrome P450. Drug Metabolism Disposition 1, 92-97 (1973).

COOPER, D.Y., SCHLEYER, H., ROSENTHAL, O.: Some chemical properties of cytochrome P-450 and its carbon monoxide compound (P-450.CO). Ann. N.Y. Acad. Sci. 174, 205-217 (1970).

CORREIA, M.A., MANNERING, G.J.: Reduced diphosphopyridine nucleotide synergism of the reduced triphosphopyridine nucleotide-dependent mixed-function oxidase system of hepatic microsomes. I. Effects of activation and inhibition of the fatty acyl coenzyme A desaturation system. Mol. Pharmacol. 9, 455-469 (1973).

DALLNER, G.: Molecular organization of the endoplasmic membranes. In: Proc. 4th Int. Cong. Pharmacol., vol. 4, pp. 70-78. Basel-Stuttgart: Schwabe, 1969.

DALLNER, G., SIEKEVITZ, P., PALADE, G.E.: Biogenesis of endoplasmic reticulum membranes. I. Structural and chemical differentiation in developing rat hepatocyte. J. Cell Biol. 30, 73-96 (1966).

DALY, J.W., JERINA, D.M., WITKOP, B.: Arene oxides and the NIH shift: The metabolism, toxicity and carcinogenicity of aromatic compounds. Experientia 28, 1129-1149 (1972).

DAVIES, D.S., GIGON, P.L., GILLETTE, J.R.: Species and sex differences in electron transport system in liver microsomes and their relationship to ethyl morphine demethylation. Life Sci. 8, 85-91 (1969).

DE KRUIFF, B., VAN GOLDE, L.M.G., VAN DEENEN, L.L.M.: Utilization of diacylglycerol species by cholinephosphotransferase, ethanolamine-phosphotransferase and diacylglycerol acyltransferase in rat liver microsomes. Biochim. Biophys. Acta 210, 425-435 (1970).

DIEHL, H., SCHADELIN, J., ULLRICH, V.: Studies on the kinetics of cytochrome P-450 reduction in rat liver microsomes. Z. Physiol. Chem. 351, 1359-1371 (1970).

DIGNAM, J.D., STROBEL, H.W.: Preparation of homogeneous NADPH-cytochrome-P450 reductase from rat liver. Biochem. Biophys. Res. Commun. 63, 845-852 (1975).

DUPPEL, W., LEBEAULT, J.M., COON, M.J.: Properties of a yeast cytochrome P450-containing enzyme system which catalyzes the hydroxylation of fatty acids, alkanes and drugs. Europ. J. Biochem. 36, 583-592 (1973).

DUS, K., KATAGIRI, M., YU, C.A., ERBES, D.L., GUNSALUS, I.C.: Chemical characterization of cytochrome P450 cam. Biochem. Biophys. Res. Commun. 40, 1423-1430 (1970).

DUS, K., LITCHFIELD, W.J., MIGUEL, A.G., VAN DER HOEVEN, A., HAUGEN, D.A., DEAN, W.L., COON, M.J.: Structural resemblance of cytochrome P450 isolated from pseudomonas putida and from rabbit liver microsomes. Biochem. Biophys. Res. Commun. 60, 15-21 (1974).

DUTTON, G.J.: Glucuronide-forming enzymes. In: Handbook of Experimental Pharmacology (Eds. B.B. BRODIE, J.R. GILLETTE, H.S. ACKERMANN), vol. XXVIII, pt.2, pp. 378-400. Berlin: Springer, 1971.

EIGEN, M.: Diffusion control in biochemical reactions. In: Quantum Statistical Mechanics in the Natural Sciences (Eds. B. KURSUNOGLU, S.L. MINTZ, S.M. WIDMAYER), pp. 37-61. New York: Plenum Press, 1974.

ELETR, S., ZAKIM, D., VESSEY, D.A.: A spin label study on the role of phospholipids in the regulation of membrane-bound microsomal enzymes. J. Mol. Biol. 78, 351-362 (1973).

ELING, T.E., DI AUGUSTINE, R.P.: A role for phospholipids in the binding and metabolism of drugs by hepatic microsomes. Biochem. J. 123, 539-549 (1971).

ERNSTER, L., JONES, L.C.: Nucleoside tri- and diphosphate activities of rat liver microsomes. J. Cell Biol. 15, 563-578 (1962).

ERNSTER, L., ORRENIUS, S.: Substrate-induced synthesis of the hydroxylating enzyme system of liver microsomes. Federation Proc. 24, 1190-1199 (1965).

ESTABROOK, R.W., BARON, J., PETERSON, J., ISHIMURA, Y.: Oxygenated cytochrome P450 as an intermediate in hydroxylation reactions. In: Biological Hydroxylation Mechanisms (Eds. G.S. BOYD, R.M.S. SMELLIE), pp. 159-185. London-New York: Academic Press, 1972.

ESTABROOᵏ, R.W., COHEN, B.: Organization of the microsomal electron transport system. In: Microsomes and Drug Oxidation (Eds. J.R. GILLETTE, A.H. CONNEY, G.J. COSMIDES, R.W. ESTABROOK, J.R. FOUTS, G.J. MANNERING), pp. 95-109. New York-London: Academic Press, 1969.

ESTABROOK, R.W., FRANKLIN, M.R., COHEN, B., SHIGAMATZU, A., HILDEBRANDT, A.G.: Influence of hepatic microsomal mixed function oxidation reactions on cellular metabolic control. Metabolism 20, 187-199 (1971).

FOUTS, J.R., ROGERS, L.A.: Morphological changes in the liver accompanying stimulation of microsomal drug metabolizing enzyme activity by phenobarbital, chlordane, benzpyrene or methylcholanthrene in rats. J. Pharmacol. Exp. Ther. 147, 112-119 (1965).

FRANZ, W., KRISCH, K.: Carboxylesterase aus Schweinenierenmikrosomen. I. Isolierung, Eigenschaften und Substratspezifitat. Z. Physiol. Chem. 349, 575-587 (1968a).

FRANZ, W., KRISCH, K.: Acylgruppenubertragung auf aromatische Amine durch Carboxylesterasen. Z. Physiol. Chem. 349, 1413-1422 (1968b).

FROMMER, U., ULLRICH, V., STAUDINGER, H.J.: Hydroxylation of aliphatic compounds by liver microsomes. I. The distribution pattern of isomeric alcohols. Z. Physiol. Chem. 351, 903-912 (1970).

GARFINKEL, D.: Studies on pig liver microsomes. I. Enzymic and pigment composition of different microsomal fractions. Arch. Biochem. Biophys. 77, 493-509 (1958).

GARLAND, R.C., CORI, C.F., CHANG, H.F.W.: Relipidation of phospholipid-depleted microsomal particles with high glucose 6-phosphatase activity. Proc. Nat. Acad. Sci. 71, 3805-3809 (1974).

GAUDETTE, L.E., BRODIE, B.B.: Relationship between the lipid solubility of drugs and their oxidation by liver microsomes. Biochem.
Pharmacol. 2, 89-96 (1959).

GIGON, P.L., GRAM, T.E., GILLETTE, J.R.: Studies on the rate of reduction of hepatic microsomal cytochrome P450 by reduced nicotinamide
adenine dinucleotide phosphate: Effect of drug substrates. Mol.
Pharmacol. 5, 109-122 (1969).

GILLETTE, J.R.: Effect of various inducers on electron transport
system associated with drug metabolism by liver microsomes.
Metabolism 20, 215-227 (1971).

GILLETTE, J.R.: Overview of drug-protein binding. Ann. N.Y. Acad. Sci.
226, 6-17 (1973).

GILLETTE, J.R., DAVIS, D.C., SASAME, H.A.: Cytochrome P450 and its
role in drug metabolism. Ann. Rev. Pharmacol. 12, 57-84 (1972).

GILLETTE, J.R., KAMM, J.J.: The enzymatic formation of sufoxides: The
oxidation of chlorpromazine and 4,4'-diaminophenyl sulfide by guinea
pig liver microsomes. J. Pharmacol. Exp. Ther. 130, 262-267 (1960).

GILLETTE, J.R., MITCHELL, J.R., BRODIE, B.B.: Biochemical mechanisms
of drug toxicity. Ann. Rev. Pharmacol. 1974, 271-288 (1974).

GLAUMANN, H.: Chemical and enzymatic composition of microsomal sub-
fractions from rat liver after treatment with phenobarbital and
3-methylcholanthrene. Chem. Biol. Interactions 2, 369-380 (1970).

GLAUMANN, H., JAKOBSSON, S.: Localization of drug metabolic activities
in liver and kidney microsomes. In: Proc. 4th Int. Cong. Pharmacol.,
vol. 4, pp. 79-86. Basel-Stuttgart: Schwabe, 1969.

GRAM, T.E., SCHROEDER, D.H., DAVIS, D.C., REAGAN, R.L., GUARINO, A.M.:
Enzymic and biochemical composition of smooth and rough microsomal
membranes from monkey, guinea pig and mouse liver. Biochem.
Pharmacol. 20, 1371-1381 (1971).

GREENZAID, P., JENCKS, W.P.: Pig liver esterase. Reactions with alcohols, structure-reactivity correlations and the acyl-enzyme intermediate. Biochemistry 10, 1210-1222 (1971).

GREIM, H.: Synthesesteigerung und Abbauhemmung bei der Vermehrung der
mikrosomalen Cytochrome P450 und b5 durch Phenobarbital. Naunyn-
Schmiedeberg's Arch. Pharmak. 266, 261-275 (1970).

GRIFFIN, B.W., PETERSON, J.A.: Camphor binding by pseudomonas putida
cytochrome P450. Kinetics and thermodynamics of the reaction.
Biochemistry 11, 4740-4746 (1972).

GUNSALUS, I.C.: Discussion remark of J. BARON, A.G. HILDEBRANDT, J.A.
PETERSON, R.W. ESTABROOK: The role of oxygenated cytochrome P-450
and of cytochrome b5 in hepatic microsomal drug oxidations. In:
Microsomes and Drug Oxidations (Eds. R.W. ESTABROOK, J.R. GILLETTE,
K.C. LEIBMAN), pp. 134-136. Baltimore: Williams & Wilkins Co., 1972.

GUNSALUS, I.C., LIPSCOMB, J.D., MARSHALL, V., FRAUENFELDER, H.,
GREENBAUM, E., MUNCK, E.: Structure and reaction of oxygenase active
centres: Cytochrome P-450 and iron sulphur proteins. In: Biological
Hydroxylation Mechanisms (Eds. G.S. BOYD, R.M.S. SMELLIE), pp. 135-
157. London: Academic Press, 1972.

HAMMES, G.G., TALLMANN, D.E.: A nuclear magnetic resonance study of
the interaction of L-epinephrine with phospholipid vesicles.
Biochim. Biophys. Acta 233, 17-25 (1971).

HAWKINS, H.C., FREEDMAN, R.B.: Fluorescence studies of drug and cation
interactions with microsomal membranes. FEBS Letters 31, 301-307
(1973).

HAYAISHI, O.: Oxygenases: In: 6th Int. Cong. Biochem. Proc. Plenary
Sessions and Program, vol. XXXIII, pp. 31-43. Washington DC 1964.

HILDEBRANDT, A.G.: Discussion remark of: R.W. ESTABROOK, T. MATSUBARA,
J.I. MASON, J. WERRINGLOER, J. BARON: Studies on the molecular func-
tion of cytochrome P-450 during drug metabolism. Drug Metabolism
Disposition 1, 109-110 (1973).

HILDEBRANDT, H., ESTABROOK, R.W.: Evidence for the participation of
cytochrome b5 in hepatic microsomal mixed-function oxidation reactions. Arch. Biochem. Biophys. 143, 66-79 (1971).

HOLLOWAY, P.W., KATZ, J.T.: A requirement of cytochrome b5 in microsomal stearyl coenzyme A desaturation. Biochem. 11, 3689-3695 (1972).

ICHIHARA, K., KUSUNOSE, E., KUSUNOSE, M.: Reconstitution of a fatty
acid w-hydroxylation system by a solubilized kidney microsomal preparation, ferredoxin, and ferredoxin-NADP-reductase. Biochim. Biophys.
Acta 202, 560-562 (1970).

ICHIKAWA, Y., YAMANO, T.: Electron spin resonance of microsomal cytochromes. Arch. Biochem. Biophys. 121, 742-749 (1967).

JEFCOATE, C.R.E., GAYLOR, J.L., CALABRESE, R.L.: Ligand interactions
with cytochrome P450. I. Binding of primary amines. Biochemistry 8,
3455-3463 (1969).

JUNGE, W., KRISCH, K.: Current problems on the structure and classification of mammalian liver carboxylesterases. Mol. Cell. Biochemistry
1, 41-52 (1973).

KADLUBAR, F.F., ZIEGLER, D.M.: Properties of a NADH-dependent N-hydroxy
amine reductase isolated from pig liver microsomes. Arch. Biochem.
Biophys. 162, 83-92 (1974).

KASCHNITZ, R., COON, M.J.: Solubilized human liver cytochrome P450:
Phospholipid requirement in hydroxylation reactions. In: Abstr. 5th
Int. Cong. Pharmacol., p. 120. White Plains, New York: Phiebig, 1972.

KATAGIRI, M., GANGULI, B.N., GUNSALUS, I.C.: A soluble cytochrome
P-450 functional in methylene hydroxylation. J. Biol. Chem. 243,
3543-3546 (1968).

KAWALEK, J.C., LU, A.Y.H.: Reconstituted liver microsomal enzyme
system that hydroxylates drugs, other foreign compounds, and endogenous substrates. VIII. Different catalytic activities of rabbit
and rat cytochromes P-448. Mol. Pharmacol. 11, 201-210 (1975).

KAWASAKI, T., YAMASHINA, I.: Isolation and characterization of glycopeptides from rough and smooth microsomes of rat liver. J. Biochem.
74, 639-647 (1973).

KHANDWALA, A.S., KASPER, C.B.: Membrane structure: The reactivity of
tryptophan, tyrosine and lysine in proteins of the microsomal membrane. Biochim. Biophys. Acta 233, 348-357 (1971).

KLINGENBERG, M.: Pigments of rat liver microsomes. Arch. Biochem.
Biophys. 75, 376-386 (1958).

KRISCH, K.: Carboxylic ester hydrolases. In: The Enzymes (Ed. P.D.
BOYER), vol. V, pp. 43-69. New York-London: Academic Press, 1971.

KUNERT, M., HEYMANN, E.: The equivalent weight of pig liver carboxylesterase and the esterase content of microsomes. FEBS Letters 49,
292-296 (1975).

LAITINEN, M., LANG, M., HÄNNINEN, O.: Changes in the protein-lipid
interaction in rat liver microsomes after pretreatment of rat rat
with barbiturates and polycyclic hydrocarbons. Int. J. Biochem. 5,
747-751 (1974).

LANDRISCINA, C., GNONI, G.V., QUAGLIARIELLO, E.: Mechanisms of fatty acid synthesis in rat liver microsomes. Biochim. Biophys. Acta 202, 405-414 (1970).

LAYNE, D.S.: New metabolic conjugates of steroids. In; Metabolic Conjugation and Metabolic Hydrolysis (Ed. W.H. FISHMAN), pp. 22-52. New York: Academic Press, 1970.

LEBEAULT, J.M., LODE, E.T., COON, M.J.: Fatty acid and hydrocarbon hydroxylation in yeast: Role of cytochrome P450 in candida tropicalis. Biochem. Biophys. Res. Commun. 42, 413-419 (1971).

LEVIN, W., RYAN, D., WEST, S., LU, A.Y.H.: Preparation of partially purified lipid-depleted cytochrome P-450 and reduced nicotinamide adenine dinucleotide phosphate-cytochrome c reductase from rat liver microsomes. J. Biol. Chem. 249, 1747-1754 (1974).

LIPSCOMB, J.D., GUNSALUS, I.C.: Structural aspects of the active site of cytochrome P450 cam. Drug Metabolism Disposition 1, 1-5 (1973).

LU, A.Y.H., COON, M.J.: Role of hemoprotein P-450 in fatty acid w-hydroxylation in a soluble enzyme system from liver microsomes. J. Biol. Chem. 243, 1331-1332 (1968).

LU, A.Y.H., JUNK, K.W., COON, M.J.: Resolution of the cytochrome P450-containing w-hydroxylation system of liver microsomes into three components. J. Biol. Chem. 244, 3714-3721 (1969a).

LU, A.Y.H., STROBEL, H.W., COON, M.J.: Hydroxylation of benzphetamine and other drugs by a solubilized form of cytochrome P-450 from liver microsomes: Lipid requirement for drug demethylation. Biochem. Biophys. Res. Commun. 36, 545-551 (1969b).

LU, A.Y.H., WEST, S.B., VORE, M., RYAN, D., LEVIN, W.: Role of cytochrome b5 in hydroxylation by a reconstituted cytochrome P450-containing system. J. Biol. Chem. 249, 6701-6709 (1974).

MAGEE, P.N., SCHOENTAL, R.: Carcinogenesis by nitroso compounds. Brit. Med. Bull. 20, 102-106 (1964).

MASTERS, B.S.S., ZIEGLER, D.M.: The distinct nature and function of NADPH-cytochrome c reductase and amine oxidase of porcine liver microsomes. Arch. Biochem. Biophys. 145, 358-364 (1971).

MATHEWS, F.S., ARGOS, P., LEVINE, M.: The structure of cytochrome b5 at 2,0 Å resolution. Cold Spring Harbor Symp. Quant. Biol. 36, 387-395 (1971).

METCALFE, J.C.: The dynamic properties of lipid molecules. In: Functional Linkage in Biomolecular System (Eds. F.O. SCHMITT, D.M. SCHNEIDER, D.M. CROTHERS), pp. 90-101. New York: Raven Press, 1975.

MIETTINEN, T.A., LESKINEN, E.: Glucuronic acid pathway. In: Metabolic Conjugation and Metabolic Hydrolysis (Ed. W.H. FISHMAN), pp. 158-237. New York: Academic Press, 1970.

MILLER, E.C., MILLER, J.A.: Mechanisms of chemical carcinogenesis: Nature of proximate carcinogens and interactions with macromolecules. Pharmacol. Rev. 18, 805-838 (1966).

MITCHELL, J.R., JOLLOW, D.J., GILLETTE, J.R., BRODIE, B.B.: Drug metabolism as a cause of drug toxicity. Drug Metabolism Disposition 1, 418-423 (1973).

MOULÉ, Y.: Biochemical characterization of the components of the endoplasmic reticulum in rat liver cell. In: Structure and Function of the Endoplasmic Reticulum in Animal Cells (Ed. F.C. GRAN), pp. 1-12. London-New York-Oslo: Universitets Forlaget, 1968.

NILSSON, R., PETTERSSON, E., DALLNER, G.: Permeability properties of rat liver endoplasmic reticulum. FEBS Letters 15, 85-88 (1971).

NOVAK, R.F., SWIFT, T.J.: Barbiturate interaction with phosphatidyl-choline. Proc. Nat. Acad. Sci. 69, 640-642 (1972).

OESCH, F.: Mammalian epoxide hydrases: Inducible enzymes catalyzing the inactivation of carcinogenic and cytotoxic metabolites derived from aromatic and olefinic compounds. Xenobiotica 3, 305-340 (1972).

ORRENIUS, S.: Induction of the drug-hydroxylating enzyme system of liver microsomes. J. Cell Biol. 26, 725-733 (1965).

ORRENIUS, S., DAS, M., GNOSSPELIUS, Y.: Overall biochemical effects of drug induction on liver microsomes. In: Microsomes and Drug Oxidation (Eds. J.R. GILLETTE, A.H. CONNEY, G.J. COSMIDES, R.W. ESTABROOK, J.R. FOUTS, G.J. MANNERING), pp. 251-277. New York-London: Academic Press, 1969.

ORRENIUS, S., ERICSSON, J.L., ERNSTER, L.: Phenobarbital-induced synthesis of the microsomal drug metabolizing enzyme system and its relationship to the proliferation of endoplasmic membranes. A morphological and biochemical study. J. Cell Biol. 25, 627-639 (1965).

ORRENIUS, S., ERNSTER, L.: Phenobarbital-induced synthesis of the oxidative demethylating enzymes of rat liver microsomes. Biochem. Biophys. Res. Commun. 16, 60-65 (1964).

ORRENIUS, S., WILSON, B.J., VON BAHR, C., SCHENKMAN, J.B.: On the significance of drug-induced spectral changes in liver microsomes. In: Biological Hydroxylation Mechanisms (Eds. G.S. BOYD, R.M.S. SMELLIE), pp. 55-77. London-New York: Academic Press, 1972.

OSHINO, N., IMAI, Y., SATO, R.: A function of cytochrome b5 in fatty acid desaturation by rat liver microsomes. J. Biochem. (Tokyo) 69, 155-162 (1971).

OSHINO, N., OMURA, T.: Immunochemical evidence for the participation of cytochrome b5 in microsomal stearyl-CoA desaturation reaction. Arch. Biochem. Biophys. 157, 395-404 (1973).

PACHE, W., CHAPMAN, D.: Interaction of antibiotics with membranes: Chlorothricin. Biochim. Biophys. Acta 255, 348-357 (1972).

PALADE, G.E., SIEKEVITZ, P.: Liver microsomes. An integrated morphological and biochemical study. J. Biophys. Biochem. Cytol. 2, 171-200 (1956).

PEISACH, J., APPLEBY, C.A., BLUMBERG, W.E.: Electron paramagnetic resonance and temperature dependent spin state studies of ferric cytochrome P450 from rhizobium iaponicum. Arch. Biochem. Biophys. 150, 725-732 (1972).

PESTKA, S.: Inhibitors of ribosome functions. Ann. Rev. Biochem. 40, 697-710 (1971).

PETERS, T., Jr., FLEISCHER, B., FLEISCHER, S.: The biosynthesis of rat serum albumin. IV. apparent passage of albumin through the golgi apparatus during secretion. J. Biol. Chem. 246, 240-244 (1971).

PETERSON, J.A.: Camphor binding by pseudomonas putida cytochrome P450. Arch. Biochem. Biophys. 144, 678-693 (1971).

POULSEN, L.L., HYSLOP, R.M., ZIEGLER, D.M.: S-oxidation of thioureylenes catalyzed by a microsomal flavoprotein mixed function oxidase. Biochem. Pharmacol. 23, 3431-3440 (1974).

PUUKKA, R., LAITINEN, M., VAINIO, H., HÄNNINEN, O.: Hepatic UDP-glucuronosyltransferase: Partial purification after 3-methyl-cholanthrene pretreatment of the rats. Int. J. Biochem. 6, 267-270 (1975).

REID, W.D., KRISHNA, G.: Centrolobular hepatic necrosis related to covalent binding of metabolites of halogenated aromatic hydrocarbons. Exp. Mol. Pathol. 18, 80-99 (1973).

REMMER, H.: The induction of hydroxylating enzymes by drugs. In: Biochemical Aspects of Antimetabolites and of Drug Hydroxylation (Ed. D. SHUGAR), vol. 16, pp. 125-141. London-New York: Academic Press, 1969.

REMMER, H., MERKER, H.J.: Enzyminduktion und Vermehrung von endoplasmatischem Reticulum in der Leberzelle wahrend der Behandlung mit Phenobarbital (Luminal). Klin. Wschr. 41, 276-283 (1963).

REMMER, H., SCHENKMAN, J., ESTABROOK, R.W., SASAME, H., GILLETTE, J., NARASIMHULU, S., COOPER, D.Y., ROSENTHAL, O.: Drug interaction with hepatic microsomal cytochrome. Mol. Pharmacol. 2, 187-190 (1966).

REMMER, H., SCHENKMAN, J.B., GREIM, H.: Spectral investigations on cytochrome P450. In: Microsomes and Drug Oxidation (Eds. J.R. GILLETTE, A.H. CONNEY, G.J. COSMIDES, R.W. ESTABROOK, J.R. FOUTS, G.J. MANNERING), pp. 371-386. New York: Academic Press, 1969.

REYNOLDS, E.S.: Liver parenchymal cell injury. IV. Pattern of incorporation of carbon and chlorine from carbon tetrachloride into chemical constituents of liver in vivo. J. Pharmacol. Exp. Ther. 155, 117-126 (1967).

RICH, P.R.: Cytochrome P-450 of higher plants: Its relation to other systems and reactivity. Biochem. Soc. Trans. 3, 980-981 (1975).

ROGERS, M.J., STRITTMATTER, P.: The binding of reduced nicotinamide adenine dinucleotide-cytochrome b5 reductase to hepatic microsomes. J. Biol. Chem. 249, 5565-5569 (1974).

ROUSER, G., NELSON, G.J., FLEISCHER, S., SIMON, G.: Lipid composition of animal cell membranes, organelles and organs. In: Biological Membranes. Physical Fact and Function (Ed.D. CHAPMAN), vol. I pp. 5-69. London-New York: Academic Press, 1968.

RYAN, D., LU, A.Y.H., WEST, S., LEVIN, W.: Multiple forms of cytochrome P450 in phenobarbital and 3-methylcholanthrene-treated rats. Separation and spectral properties. J. Biol. Chem. 250, 2157-2163 (1975a).

RYAN, D., LU, A.Y.H., KAWALEK, J., WEST, S.B., LEVIN, W.: Highly purified cytochrome P448 and P450 from rat liver microsomes. Biochem. Biophys. Res. Commun. 64, 1134-1141 (1975b).

SABATINI, D.D., TASHIRO, Y., PALADE, G.E.: On the attachment of ribosomes to microsomal membranes. J. Mol. Biol. 19, 503-524 (1966).

SASAME, H.A., MITCHELL, J.R., THORGEIRSSON, S., GILLETTE, J.R.: Relationship between NADH and NADPH oxidation during drug metabolism. Drug Metabolism Disposition 1, 150-155 (1973).

SATO, R., NISHIBAYASHI, H., ITO, A.: Characterization of two hemoproteins of liver microsomes. In: Microsomes and Drug Oxidation (Eds. J.R. GILLETTE, A.H. CONNEY, G.J. COSMIDES, R.W. ESTABROOK, J.R. FOUTS, G.J. MANNERING), pp. 111-132. New York: Academic Press, 1969.

SCHENKMAN, J.B.: Studies on the nature of the type I and type II spectral changes in liver microsomes. Biochemistry 9, 2081-2091 (1970).

SCHENKMAN, J.B., CINTI, D.L., ORRENIUS, S., MOLDEUS, P., KASCHNITZ, R.: The nature of reverse type I (modified type II) spectral changes in liver microsomes. Biochemistry 11, 4243-4251 (1972).

SCHENKMAN, J.B., REMMER, H., ESTABROOK, R.W.: Spectral studies of drug interaction with hepatic microsomal cytochrome. Mol. Pharmacol. 3, 113-123 (1967).

SCHENKMAN, J.B., SATO, R.: The relationship between the pH-induced spectral change in ferriprotoheme and the substrate induced spectral change of the hepatic microsomal mixed function oxidase. Mol. Pharmacol. 4, 613-620 (1968).

SCHLEYER, H., COOPER, D.H., ROSENTHAL, O.: The hemeprotein P450 in steroid hydroxylation. Ann. N.Y. Acad. Sci. 222, 102-117 (1973).

SCHULZE, H.U., STAUDINGER, H.J.: Zur Struktur des endoplasmatischen Retikulums der Rattenleberzelle: Korrelation von morphometrischen und biochemischen Me werten. Hoppe Seyler's Z. Physiol. Chem. 352, 1675-1680 (1971).

SCHUSTER, I., FLESCHURZ, C., HELM, I.: On the interaction of a lipophilic drug with different sites of rat-liver microsomes. Equilibrium studies with a substituted pleuromutilin. Europ. J. Biochem. 51, 511-519 (1975).

SCHUSTER, I., FLESCHURZ, C., HELM, I.: Kinetics of tiamutin[R] interaction with cytochrome P450 from rabbit liver in microsomes and drug oxidation (Eds. A.H. CONNEY, R.W. ESTABROOK, A.G. HILDEBRANDT, V. ULLRICH). New York: Pergamon Press, 1977 (in press).

SCHUSTER, I., HELM, I., FLESCHURZ, C.: The effect of charcoal treatment on microsomal cytochrome P450. FEBS Letters (1977, in press).

SEEMAN, P.: The membrane actions of anaesthethics and tranquilizers. Pharmacol. Rev. 24, 583-655 (1972).

SHANK, R.C.: Toxicology of N-nitroso compounds. Toxicol. Appl. Pharmacol. 31, 361-368 (1975).

SHARROCK, M., MÜNCK, E., DEBRUNNER, P.G., MARSHALL, V., LIPSCOMB, J.D., GUNSALUS, I.C.: Mössbauer studies of cytochrome P450 cam. Biochemistry 12, 258-265 (1973).

SHIRES, T.K., McLAUGHLIN, C.M., PITOT, M.C.: The selectivity and stoichiometry of membrane binding sites for polyribosomes, ribosomes and ribosomal subunits in vitro. Biochem. J. 146, 513-526 (1975).

SIEBERT, G.: Biochemie der Zellstrukturen. In: Handbuch der allgemeinen Pathologie (Eds. H.W. ALTMANN, F. BÜCHNER, H. COTTIER, G. HOLLE, E. LETTERER, W. MASSHOFF, H. MEESEN, F. ROULET, G. SEIFERT, G. SIEBERT, A. STUDER), vol. 2, pt. 5, pp. 140-153. Berlin: Springer, 1968.

SPATZ, L., STRITTMATTER, P.: A form of cytochrome b5 that contains an additional hydrophobic sequence of 40 amino acid residues. Proc. Nat. Acad. Sci. 68, 1042-1046 (1971).

SPATZ, L., STRITTMATTER, P.: A form of reduced nicotinamide adenine dinucleotide-cytochrome b5 reductase containing both the catalytic site and an additional hydrophobic membrane=binding segment. J. Biol. Chem. 248, 793-799 (1973).

STERN, J.O., PEISACH, J. BLUMBERG, W.E., LU, A.Y.H., LEVIN, W.: A low temperature EPR study of partially purified soluble ferric cytochromes P450 and P448 from rat liver microsomes. Arch. Biochem. Biophys. 156, 404-413 (1973).

STIER, A., SACKMANN, E.: Spin labels as enzyme substrates. Heterogeneous lipid distribution in liver microsomal membranes. Biochim. Biophys. Acta 311, 400-408 (1973).

STOOPS, J.K., HORGAN, D.J., RUNNEGAR, M.T.C., DE JERSEY, J., WEBB, E.C., ZERNER, B.: Carboxylesterases (EC 3.1.1.). Kinetic studies on carboxylesterases. Biochemistry 8, 2026-2033 (1969).

STRITTMATTER, P., VELICK, S.F.: The purification and properties of microsomal cytochrome reductase. J. Biol. Chem. 228, 785-799 (1957).

STROBEL, H.W., LU, A.Y.H., HEIDEMA, J., COON, M.J.: Phosphatidylcholine requirement in the enzymatic reduction of hemoprotein P450 and in fatty acid, hydrocarbon and drug hydroxylation. J. Biol. Chem. 245, 4851-4854 (1970).

SVENSSON, H., DALLNER, G., ERNSTER, L.: Investigation of specificity
in membrane breakage occurring during sonication of rough microsomal
membranes. Biochim. Biophys. Acta 274, 447-461 (1972).

TRÄUBLE, H., EIBL, H.J.: Cooperative structural changes in lipid bi-
layers. In: Functional Linkage in Biomolecular Systems (Eds. F.O.
SCHMITT, D.M. SCHNEIDER, D.M. CROTHERS), pp. 59-90. New York: Raven
Press, 1975.

TSAI, R.L., YU, C.A., GUNSALUS, I.C., PEISACH, J., BLUMBERG, W.,
ORME-JOHNSON, W.H., BEINERT, H.: Spin state changes in cytochrome
P-450 cam. on binding of specific substrates. Proc. Nat. Acad. Sci.
66, 1157-1163 (1970).

TYSON, C.A., LIPSCOMB, J.D., GUNSALUS, I.C.: The role of putidaredoxin
and P450 cam. in methylene hydroxylation. J. Biol. Chem. 247, 5777-
5784 (1972).

UEHLEKE, H.: Toxicological aspects of the N-hydroxylation of aromatic
amines. Naunyn Schmiedeberg's Arch. Exp. Path. Pharmak. 263, 106-
120 (1969).

UEHLEKE, H.: N-hydroxylation. Xenobiotica 1, 327-338 (1971).

ULLRICH, V.: On the hydroxylation of cyclohexane in rat liver microsomes. Z. Physiol. Chem. 350, 357-365 (1969).

VAN DER HOEVEN, T.A., COON, M.J.: Preparation and properties of partially purified cytochrome P450 and reduced nicotinamide adenine
dinucleotide phosphate-cytochrome P-450 reductase from rat liver
microsomes. J. Biol. Chem. 249, 6302-6310 (1974).

VAZQUEZ, D., BARBACID, M., FERNANDEZ-MUÑOZ, R.: Antibiotic action on
the ribosomal peptidyltransferase center. Topics Infectious Diseases
1, 193-216 (1975).

WATERMAN, M.R., ULLRICH, V., ESTABROOK, R.W.: Effect of substrate on
the spin state of cytochrome P450 in hepatic microsomes. Arch.
Biochem. Biophys. 155, 355-360 (1973).

WEISBURGER, J.H., WEISBURGER, E.K.: Biochemical formation and pharmacological, toxicological and pathological properties of hydroxylamines and hydroxamic acids. Pharmacol. Rev. 25, 1-66 (1973).

WICKRAMASINGHE, R.H.: Biological aspects of cytochrome P450 and associated hydroxylation reactions. Enzyme 19, 348-376 (1975).

WILLIAMS, R.T.: The biogenesis of conjugation and detoxication products. In: Biogenesis of Natural Compounds (Ed. P. BERNFELD), 2nd
ed., pp. 589-639. New York: Pergamon Press, 1967.

YAMAMOTO, S., BLOCH, K.: Enzymes catalyzing the transformation of
squalene to lanosterol. Proc. Biochem. Soc. Biochem. J. 113, 19-20P
(1969).

YANG, C.S.: The association between cytochrome P450 and NADPH cytochrome P450 reductase in microsomal membrane. FEBS Letters 54, 61-
64 (1975).

YU, C.A., GUNSALUS, I.C.: Cytochrome P450 cam. II. Interconversion
with P420. J. Biol. Chem. 249, 102-106 (1974).

YU, C.A., GUNSALUS, I.C., KATAGIRI, M., SUHARA, K., TAKEMORI, S.:
Cytochrome P450 cam. I. Crystallization and properties. J. Biol.
Chem. 249, 94-101 (1974).

ZAKIM, D., GOLDENBERG, J., VESSEY, D.A.: Effects of metals on the
properties of hepatic microsomal uridine diphosphate glucuronyl-
transferase. Biochemistry 12, 4068-4074 (1973a).

ZAKIM, D., GOLDENBERG, J., VESSEY, D.A.: Influence of membrane lipids
on the regulatory properties of UDP-glucuronyl transferase. Europ.
J. Biochem. 38, 59-63 (1973b).

ZIEGLER, D.M., McKEE, E.M., POULSEN, L.L.: Microsomal flavoprotein-
  catalyzed N-oxidation of arylamides. Drug Metabolism Disposition 1,
  314-321 (1973).
ZIEGLER, D.M., MITCHELL, C.H.: Microsomal oxidase. IV. properties of
  a mixed function amine oxidase isolated from pig liver microsomes.
  Arch. Biochem. Biophys. 150, 116-125 (1972).
ZIEGLER, D.M., PETTIT, F.H.: Microsomal oxidases. I. The isolation and
  dialkylarylamine oxygenase activity of pork liver microsomes.
  Biochemistry 5, 2932-2938 (1966).
ZIEGLER, D.M., POULSEN, L.L., McKEE, E.M.: Interaction of primary
  amines with a mixed function amine oxidase isolated from pig liver
  microsomes. Xenobiotica 1, 523-531 (1971).

The Pathogenesis of Experimental Acute Renal Failure: The Role of Membrane Dysfunction

Walter Flamenbaum, John H. Schwartz, Robert J. Hamburger,
and James S. Kaufman

I. Introduction

The mechanism responsible for the pathogenesis of acute renal failure,
a disease characterized by the sudden loss of renal function, continues to be the subject of controversy. The maintained high morbidity
and mortality, despite numerous clinical advances in its treatment as
well as the renewed scientific interest in its pathogenesis, suggests
that a detailed knowledge of the pathophysiologic mechanisms will be
required before significant progress in the prevention and/or treatment of acute renal failure will be forthcoming. The controversy appears to have persisted, despite the availability of sophisticated
investigative techniques, because of the lack of a single experimental model or a single pathophysiologic mechanism that best explains
the observed abnormalities in renal function.

The approach we have utilized to study the pathophysiologic basis for
acute renal failure has been to choose a single experimental model
that approximates the functional abnormalities observed in clinical
acute renal failure and to examine it in detail using a variety of
experimental approaches. The model chosen for study was uranyl
nitrate-induced acute renal failure. Based on our investigations of
this model of experimental acute renal failure, the following schema
for the pathophysiologic basis of acute renal failure has been proposed:

1. The initial event is an alteration in tubular epithelial (membrane) function characterized by decreased fluid and electrolyte
   absorption (SCHWARTZ and FLAMENBAUM, in press; FLAMENBAUM et al.,
   1974) as a direct result of the effect of uranyl nitrate on membrane function.
2. As a result of this change in fluid reabsorption, there is an alteration in the composition of tubular fluid delivered to the
   macula densa segment of the distal nephron, manifested by an increase in tubular fluid sodium concentration (FLAMENBAUM et al.,
   1975).
3. As a direct consequence of this alteration in tubular fluid composition, as sensed by the macula densa, there is an increase in
   renin-angiotensin system activity on the local nephron level, with
   an increase in juxtaglomerular renin activity (FLAMENBAUM et al.,
   1975).
4. The increased renin-angiotensin system activity mediates a change

in renal hemodynamics, characterized by a diminution in total renal blood flow with preferential outer cortical ischemia, resulting in a progressive decline in glomerular filtration rate
(FLAMENBAUM et al., 1972b; KLEINMAN et al., 1975).
5. The negative feedback loop of this mechanism fails to turn on,
   despite a decrease in filtered sodium load, because of the persistent effects of uranyl nitrate on membrane function (FLAMENBAUM et
   al., 1975).
Each of the steps in the schema of this model will now be considered
in detail.

II. Heavy Metal-Induced Alterations in Membrane Function

The effects of uranyl nitrate, and other heavy metals, on membrane
function were studied using *in vitro* and *in vivo* techniques.

A. The *in vitro* Preparation

The urinary bladder of the freshwater turtle (*Pseudemys scripta*), a
mesonephric derivative, characteristically resembles the distal nephron of the mammalian nephron in its capacity to reabsorb sodium, secrete hydrogen ion and respond to aldosterone and vasopressin
(STEINMETZ, 1974; NORBY, et al. 1975). Using this *in vitro* preparation, the direct effects of heavy metals on membrane function may be
studied without any secondary perturbations associated with systemic
alterations.

1. Methodology

Urinary bladders were mounted in lucite chambers and bathed on the
two sides with an identical amphibian Ringer's solution (SCHWARTZ and
FLAMENBAUM, 1975). The spontaneous potential difference across the
bladder was oriented such that the serosal side (body fluid side) was
normally positive with respect to the mucosal side (urinary surface).
This spontaneous potential difference (PD) was continuously short-
circuited by means of an automatic voltage clamp, and the short-
circuit current (SCC) was recorded potentiometrically. The transepithelial resistance was calculated from the PD and SCC, or change in
PD when a 100 μA current pulse was sent through the bladder. Bidirectional sodium fluxes were determined using $^{24}NaCl$ and $^{22}NaCl$, and
chloride fluxes were determined using $Na^{36}Cl$. All rates were expressed as microequivalents per hour per 8.0 square centimeters of
membrane exposed (μEq/h/8.0 cm^2), or converted to microamperes (μA)
in accordance with Faraday's law.

2. Direct Effects of Heavy Metals

a. Dose Response Curves for Uranyl Nitrate and HgCl$_2$

To establish the effects of heavy metals on membrane function, a dose-
response curve was determined using uranyl nitrate in concentrations

ranging from 1 x 10^{-7} to 1 x 10^{-2} M and measuring the change in SCC.
In Fig. 1, the percent inhibition of SCC at each concentration of
uranyl nitrate is presented. The inhibition of SCC increased from
12 ± 2(SEM)% with 1 x 10^{-7} M uranyl nitrate in the mucosal solution to
69 ± 6% with 1 x 10^{-4} M. At higher concentrations of uranyl nitrate,
the degree of inhibition of SCC did not increase significantly. Over
the entire concentration range of uranyl nitrate studied, there was
no significant alteration in transepithelial resistance (Fig. 1).
Based on this dose-response curve, the minimal concentration of uranyl
nitrate that reduced SCC by approximately 70% was studied in further
detail. The time course of SCC inhibition after addition of 1 x 10^{-4} M
uranyl nitrate to the mucosal solution is shown in Fig. 2.

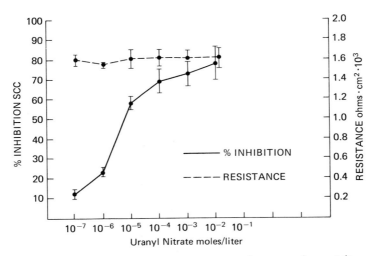

Fig. 1. Effect of increasing dose of uranyl nitrate, in mol/l, on turtle
bladder SCC (expressed as % inhibition from control) and transepithelial
resistance (ohms·cm^2·10^3). All values are expressed as mean+SEM for 6
bladders in each group

Fig. 2. Uranyl nitrate
(*UN*)-induced inhibition
of SCC as function of
time, compared with
control bladders. Val-
ues are expressed as
mean+SEM for 8 paired
bladders

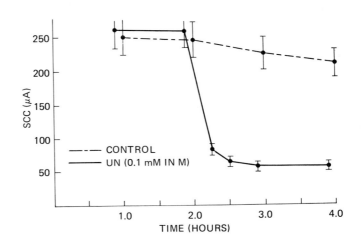

The addition of uranyl nitrate resulted in a rapid decline in the SCC, achieving a minimum value of 64 ± 9 μA within 20 min as compared to the initial value of 235 ± 2 μA. In contrast to the effect of mucosal addition, serosal addition of up to 10^{-2} M uranyl nitrate had no effect on SCC or transepithelial resistance.

In order to demonstrate that the effects of uranyl nitrate were not a singular action of this heavy metal, additional studies were performed using $HgCl_2$. Doses of $HgCl_2$ ranging from 1×10^{-8} to 1×10^{-3} M were added to the mucosal solution of prepared turtle bladders, and the dose-response curve depicted in Fig. 3 was constructed. The inhibition of SCC increased from $10 \pm 2\%$, at 1×10^{-8} M, to $81 \pm 3\%$, at 1×10^{-5} M $HgCl_2$. At higher concentrations, there was no further significant alteration in SCC. The transepithelial resistance did not change as compared to the initial control value of $1.31 \pm 0.09 \times 10^3$ ohm·cm , until 1×10^{-3} M $HgCl_2$ was present in the mucosal solution. At this higher concentration, tissue resistance fell to less than a third of the control value. Since the bladder tissue becomes opaque and extremely friable at this higher concentration of $HgCl_2$, these results are consistent with direct and irreversible tissue destruction that precludes further study. The dose of $HgCl_2$ selected for further study was 1×10^{-5} M because its effects on SCC and transepithelial resistance paralleled those observed with 1×10^{-4} M uranyl nitrate. The time course of SCC inhibition induced by the addition of 10^{-5} M $HgCl_2$ to the mucosal solution is depicted in Fig. 4. The SCC rapidly decreased and reached a minimal value within 20 min of the addition of $HgCl_2$ to the solution bathing the mucosal surface of the turtle bladder.

In contrast to the results observed with the serosal addition of uranyl nitrate, the serosal addition of 10^{-5} M $HgCl_2$ (Fig. 5) resulted

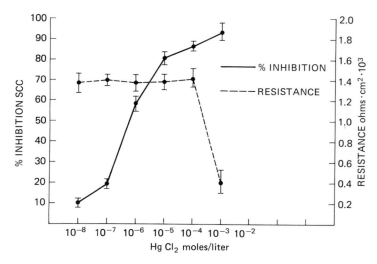

Fig. 3. Effect of increasing dose of $HgCl_2$, in mol/l, on turtle bladder SCC (expressed as % inhibition from control) and transepithelial resistance (ohms·cm^2·10^3). All values are expressed as mean+SEM for 6 bladders in each group

Fig. 4. HgCl$_2$-induced inhibition of SCC as function of time, compared with control bladders. Values are expressed as mean+SEM for 8 paired bladders

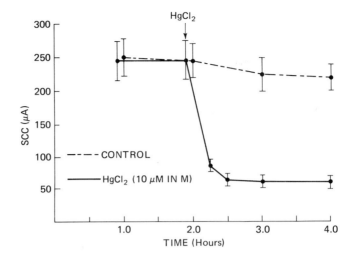

Fig. 5. Effect of serosal addition of HgCl$_2$ on SCC. HgCl$_2$, 10 μM, was added to serosal solution at 2.0 h (*solid line*), and at 3.0 h HgCl$_2$ was removed by washing or by addition of 2.0 mM dithiothreitol (*dashed line*) to both mucosal and serosal solutions. Values are expressed as mean+ SEM for 8 paired bladders

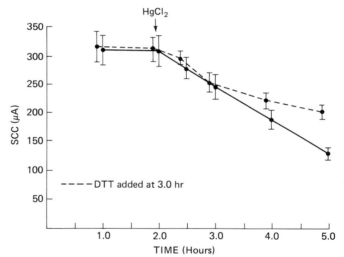

in a progressive decrease in SCC. The time course of this decrease, however, was significantly slower, -1.0 ± 0.3 A/min, than after mucosal exposure, -6.8 ± 0.2 A/min. Furthermore, a stable value for SCC was never achieved after the serosal addition of HgCl$_2$.

These results suggest that the alterations in SCC, an approximate measure of net sodium transport (STEINMETZ, 1974), probably result from an interaction of these heavy metals with constituents of the mucosal membrane. The inhibition of sodium transport apparently occurs without change in the passive ionic conductance of the transepithelial membrane, as suggested by the lack of change in the transepithelial resistance. That serosal addition has either no effect, as with uranyl nitrate, or a diminished effect with a prolonged time course, as with HgCl$_2$, on SCC is consistent with these heavy metals having a primary mucosal site of action. The results obtained with the serosal addition of HgCl$_2$ will be discussed in greater detail. below.

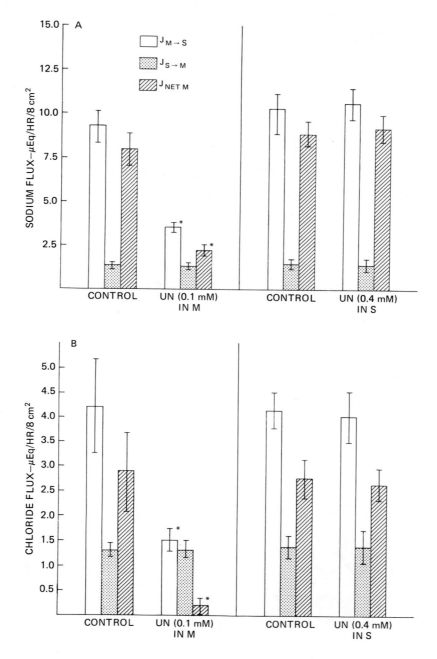

<u>Fig. 6.</u> Uranyl nitrate-induced alterations in Na⁺ fluxes (*A*) and Cl⁻ fluxes (*B*). Left-hand panels depict effect of addition of UN, 0.1 mM, to mucosal solution (*M*) of turtle bladder; right-hand panel depicts effect of addition of UN, 0.4 mM, to serosal solution (*S*) of turtle bladder. Each bar is mean±SEM for 6 studies. $J_{M \rightarrow S}$, $J_{S \rightarrow M}$, and $J_{NET \ M}$ are unidirectional fluxes and net flux from M, respectively

b. *Alterations in Electrolyte Transport*

Although the SCC is primarily determined by sodium transport, turtle bladder epithelium is also capable of electrogenic hydrogen ion transport and active coupled chloride-bicarbonate exchange (LESLIE et al., 1973; SCHWARTZ, in press). To characterize further the changes in electrolyte transport induced by uranyl nitrate and $HgCl_2$, unidirectional fluxes of $^{24}Na^+$ and $^{36}Cl^-$ were determined across paired short-circuited bladders. Prior to the addition of uranyl nitrate net Na^+ transport was 7.95 ± 0.81 $\mu Eq/h/8$ cm^2. This rate of Na^+ transport, expressed as μA, 210 ± 21 $\mu A/8$ cm^2, is not significantly different from the simultaneously measured SCC of 206 ± 15 $\mu A/8$ cm^2. The net flux of Na^+, as shown in Fig. 6, results from a 9.3 ± 0.8 $\mu Eq/h/8$ cm^2 flux of Na^+ from mucosal solution to serosal solution, and a simultaneous passive back flux of Na^+ from serosal solution to mucosal solution of 1.30 ± 0.1 $\mu Eq/h/8$ cm^2. After the addition of 10^{-5} M uranyl nitrate to the mucosal solution, the SCC decreased to 64.3 ± 2.6 $\mu A/8$ cm^2. Net Na^+ transport also decreased after addition of uranyl nitrate to a value not significantly different from the concurrently measured SCC. This decrease in Na^+ transport, as shown in Fig. 6, resulted from a decline in the active component, mucosa to serosa, of Na^+ transport without any significant alteration in the passive, serosa to mucosa, flux of Na^+. Net Cl^- transport in these same tissues was 2.88 ± 0.79 $\mu Eq/h/8$ cm^2 during the control period. After addition of uranyl nitrate to the mucosal solution, net Cl^- transport was virtually abolished. The alteration in net Cl^-, like the alteration in net Na^+ transport, resulted only from inhibition of the active, mucosa to serosa, component of net Cl^- flux (Fig. 6). Serosal addition of 4×10^{-5} M uranyl nitrate did not result in any measurable changes in the fluxes of either Na^+ or Cl^-.

Parallel studies of ion fluxes across bladders exposed to 10^{-5} M $HgCl_2$ are presented in Fig. 7. These studies demonstrate that the decrease in SCC after addition of $HgCl_2$ to the mucosal solution results from a decrease in the active component of Na^+ transport, without measurable change in passive Na^+ flux. Unlike uranyl nitrate, 10^{-5} M $HgCl_2$ had no effect on either the net flux or unidirectional fluxes of Cl^-. Ion fluxes were not examined after serosal addition of $HgCl_2$, since stable electrophysiologic parameters were never achieved over the time course required for the performance of these studies.

Several important observations should be made concerning the results of these studies on the effects of heavy metals on membrane function. First, both uranyl nitrate and $HgCl_2$ inhibit active Na^+ transport as reflected by both the changes in SCC and isotopic fluxes. These changes occur without altering the ionic conductance of the membrane as demonstrated by the constancy of the transepithelial resistance and the passive component of ionic fluxes. Lastly, these alterations in ionic fluxes and the electrophysiologic changes occur primarily after mucosal exposure to the heavy metal. These observations indicate that membrane integrity is not markedly altered, suggesting that the probable site of heavy metal-membrane interaction resides within the apical mucosal membrane of the epithelial cells in this preparation.

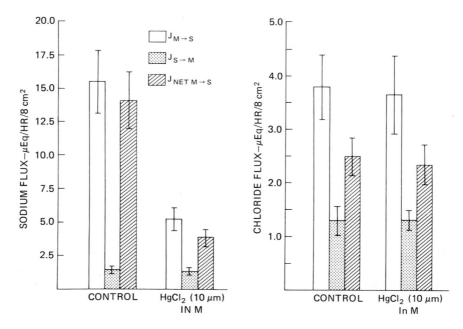

Fig. 7. Effect of HgCl$_2$ on Na$^+$ fluxes (*left-hand panel*) and Cl$^-$ fluxes (*right-hand panel*) after addition of 10 μM to mucosal solution (*M*) of 6 paired tissues. Notations for fluxes (*J*) are as indicated in Fig. 6

c. Membrane Site of Heavy Metal Action

Studies with Amphotericin B. If the site of membrane interaction with heavy metals, as proposed above, involves solely the plasma membrane of the mucosal side of the epithelial cell, then it is probable that functions carried out within intracellular sites, such as the enzyme systems providing energy for the active transport processes and the antiluminal sites of active transport, are not affected. According to this proposal, the decline in Na$^+$ observed after the addition of a heavy metal salt to the mucosal solution may alter an apical entry site for Na$^+$ rather than affect the electrogenic transport of Na$^+$ out of the cell at the antiluminal surface. Thus, after exposure to heavy metal, the intracellular pools of Na$^+$ diminish as a consequence of the heavy metal-apical membrane interaction. Although this proposal excludes direct effects of heavy metals on the antiluminal sites of active ion transport, it is possible that heavy metals may indirectly affect these sites by decreasing the magnitude of the intracellular sodium pool available for transport. In order to evaluate the possibility that a heavy metal-induced alteration in Na$^+$ entry into the apical cell membrane was responsible for the observed change in net Na$^+$ transport, a maneuver known to increase apical Na$^+$ conductance should reverse the inhibition of Na$^+$ transport. Amphotericin B has been previously demonstrated to increase the permeability of the mucosal barrier of amphibian urinary bladders to solute by interacting with membrane cholesterol (FINN, 1968; LICHTENSTEIN and LEAF, 1969;

STEINMETZ and LAWSON, 1970). An increase in Na^+ transport in response to the addition of amphotericin B to bladders previously exposed to heavy metal would lend support to this hypothesis.

In bladders exposed to uranyl nitrate, 10^{-4} M, or $HgCl_2$, 10^{-5} M, in the mucosal solution the addition of 20 $\mu g/ml$ of amphotericin B to the mucosal solution resulted in the changes in SCC depicted in Fig. 8. Within 1 hour after the addition of amphotericin B, the SCC increased to values not different from the initial control values obtained prior to heavy metal exposure. The transepithelial resistance of bladders exposed to a combination of uranyl nitrate and amphotericin B decreased to $0.95 \pm 0.07 \times 10^3$ ohm·cm^2. To demonstrate that there was an equivalency between the alterations in SCC and net Na^+ transport after the addition of both uranyl nitrate and amphotericin B, bidirectional Na^+ fluxes were also determined; the results are presented in Table 1. The increased SCC after amphotericin B addition can be explained by an equivalent increase in net Na^+ transport, in studies of both uranyl nitrate and $HgCl_2$ bladders. The stimulation of sodium transport by the addition of amphotericin B to bladders previously exposed to heavy metals suggests that the cellular capacity to transport Na^+ was not altered by direct effects of heavy metals on intracellular components of Na^+ transport. Rather, the effect obtained with amphotericin B suggests that inhibition of Na^+ transport by heavy metals is the result of an alteration in the Na^+ conductance of the apical membrane of the epithelial cell.

Studies with Dithiothreitol. Accepting, for the moment, that heavy metals have their most prominent and direct effects on the apical membrane of these epithelial cells, one may consider the potential sites within those membranes in which this interaction may occur. There is considerable evidence in the literature that one heavy metal, $HgCl_2$, is preferentially bound by sulfhydryl groups of intact tissues (ROTHSTEIN, 1959; WEBB, 1966). The binding sites for $HgCl_2$, and

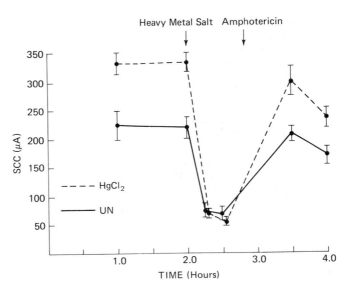

Fig. 8. Effect of amphotericin B, 20 $\mu gm/ml$, added to mucosal solution (*M*), on heavy metal-induced inhibition of SCC. Uranyl nitrate (*UN*), 0.1 mM, or $HgCl_2$, 10 μM, was added to M at 2.0 h; removed from M by washing at 2.45 h; then amphotericin B was added. Values are expressed as mean+SEM for 8 bladders

Table 1. Effect of amphotericin B on Na^+ transport and SCC across turtle
urinary bladders exposed to uranyl nitrate or mercuric chloride

Experiment	Sodium fluxes[a]			
	Mucosal to serosal flux	Serosal to mucosal flux	Net flux	SCC
Uranyl nitrate alone, 0.1 mM, in mucosal solution	4.11 ± 0.11	1.50 ± 0.14	2.67 ± 0.21	2.57 ± 0.30
Plus amphotericin B, 20 μgm/ml in mucosal solution	9.12 ± 0.21	3.04 ± 0.17	6.08 ± 0.22	5.72 ± 0.46
$HgCl_2$ alone, 10 μM, in mucosal solution	4.10 ± 0.37	1.60 ± 0.11	2.54 ± 0.27	2.68 ± 0.10
Plus amphotericin B, 20 μgm/ml, in mucosal solution	13.71 ± 1.23	3.21 ± 0.38	10.50 ± 0.80	11.00 ± 1.10

[a]Mucosal to serosal and serosal to mucosal fluxes were measured simultaneously, and are expressed as $\mu Eq/h/8cm^2$ as mean ± SEM for 6 bladders in
each study. Net Na^+ flux was not statistically different from SCC before
or after addition of amphotericin B.

perhaps uranyl nitrate, may involve sulfhydryl groups within the
apical membrane of turtle bladder epithelium. In order to evaluate
this possibility, the effects of dithiothreitol, a dithiol sugar
capable of complexing heavy metals and maintaining monothiols in a
reduced state (CLELAND, 1964), on SCC of bladders exposed to heavy
metals was determined. The effects of either uranyl nitrate (Fig. 9)
or HgCl (Fig. 10) on SCC were rapidly and completely reversed by the
addition of 2.0 mM dithiothreitol to the mucosal solution. In these
studies, after a 30 min mucosal exposure to the heavy metals, the
tissues were washed with fresh amphibian Ringer's solution to remove
free heavy metal from the bathing solution and then dithiothreitol
was added to the mucosal solution of one of the paired bladders. As
is readily apparent, simply removing free heavy metal from the mucosal
bathing solution did not materially reverse the diminution in SCC.
The addition of dithiothreitol to either the mucosal solution of control bladders or the serosal solutions of heavy metal-treated bladders
did not result in any significant alteration in SCC.

The rapid reversibility of the effect of $HgCl_2$ on SCC by dithiothreitol
is highly consistent with a role for sulfhydryl-mercury interaction in
the apical membrane. Since dithiothreitol also reverses the inhibition of SCC induced by uranyl nitrate, the binding sites for uranyl
ions may also involve sulfhydryl groups (ROTHSTEIN, 1959; BARRON et
al., 1948). To ascertain if uranyl ions can complex with sulfhydryl
groups, the ability of dithiol sugars, or cysteine, to solubilize
uranyl-hydroxyl precipitates was examined. Both dithiothreitol and
dithioerythritol, as well as cysteine, solubilize these precipitates.

Fig. 9. Effect of dithiothreitol (*DTT*), 2 mM, added to mucosal solution (*M*) on uranyl nitrate (*UN*)-induced inhibition of SCC. UN, 0.1 mM, was added to M at 2.0 h; removed at 2.3 h by washing; and, in one of each pair of bladders DTT was added (*solid line*). Values are expressed as mean+ SEM for 8 paired bladders

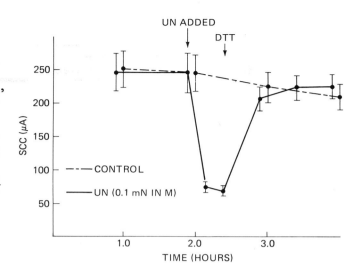

Fig. 10. Effect of dithiothreitol (*DTT*), 2 mM, added to mucosal solution (*M*) on HgCl$_2$-induced inhibition of SCC. Values are expressed as mean+SEM for 8 paired tissues

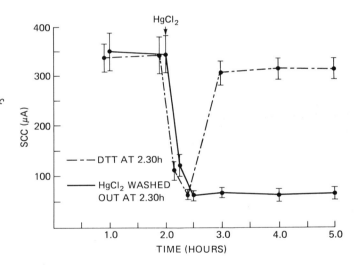

In contrast, threitol, erythritol, and cystine did not solubilize the precipitates, suggesting that sulfhydryl groups can react with uranyl ions. Thus, the reversal of uranyl nitrate-induced inhibition of SCC by dithiothreitol is consistent with the suggestion that uranyl ions may react with apical membrane sulfhydryl groups.

Interpretation of Heavy Metal Action in vitro. Any proposal to account for the inhibitory actions of uranyl nitrate or HgCl$_2$ on ion transport across this epithelial preparation must be consistent with the following observations:

1. Inhibition begins rapidly after mucosal addition, but not after serosal addition of the heavy metals.

2. There are no changes in the transepithelial electrical resistance, or change in the passive ionic conductances for Na^+ or Cl^-, after the addition of the heavy metals to the epithelium.
3. Dithiothreitol reverses heavy metal inhibition of SCC induced by heavy metals only after mucosal addition.
4. Amphotericin B also reverses the inhibition of Na^+ transport induced by heavy metals.
5. It has been previously observed that acidification of the mucosal solution also results in a similar inhibition of Na^+ transport (STEINMETZ and LAWSON, 1971).
6. Studies of the electrical profile across the turtle bladder suggest that the active step for Na^+ transport is located in the serosal (antiluminal, basilateral) membrane (HIRSCHORN and FRAZIER, 1971).

An interpretation that may account for these observations would be that heavy metals alter specific Na^+ entry sites in the apical membrane, which are either in series with, or functionally coupled to, the active transport site in the antiluminal membrane. Sites for the passive permeation of Na^+, Cl^-, and other electrolytes across the epithelium are not affected by this heavy metal-apical membrane interaction. The constancy of the passive permeability after heavy metal exposure, in part, accounts for the lack of measurable change in transepithelial electrical resistance. The prompt reversal of heavy metal inhibition of SCC by either mucosal addition of dithiothreitol, in contrast to its lack of effect when added to the serosal solution, or mucosal addition of amphotericin B further support this concept. Dithiothreitol action, by forming more stable complexes with these heavy metals than with the proposed sulfhydryl reactive sites within the apical membrane, or by regenerating sulfhydryl groups within the membrane, thereby restores the specific Na^+ entry sites to an equivalent of their original functional state. Amphotericin B may further alter the apical membrane so that Na^+ can enter through alternate pathways, to gain access to the intracellular transportable Na^+ pool. The resultant increment in the intracellular Na^+ pool restores Na^+ transport to its pre-inhibited state.

STEINMETZ and LAWSON (1971) proposed a similar interpretation to account for the inhibitory effect on Na^+ transport of acidifying the mucosal solution. In their study, lowering the pH of the mucosal solution by addition of exogenous acid, which did not result in a change in tissue (cellular) pH, resulted in a decrease in both SCC and mucosa to serosa (active component) flux of Na^+ without a decrease in passive flux. Furthermore, after active Na^+ transport was inhibited by the addition of ouabain, luminal acidification did not result in a decrease in the mucosa to serosa Na^+ flux. The proposed specific Na^+ entry sites in the apical membrane are, thus, pH and heavy metal sensitive. The sites for passive permeation are pH and heavy metal insensitive. Because the specific entry sites may be in series with the active transport sites, an alteration in one set of sites may, in some manner, affect the function of the other sites. The demonstration by REUSS and FINN (1975) that changes in the potential difference across the apical membrane induced by changing the composition of the mucosal solution or by the addition of amiloride resulted in an almost simultaneous change in the potential difference across the antiluminal membrane confirms, in part, the present proposal that heavy metal interaction with specific Na^+ entry sites in the luminal membrane may secondarily affect the active transport sites in the antiluminal membrane.

Another, less likely proposal not requiring the postulate of specific Na^+ entry sites in the apical membrane would be that heavy metals decrease Na^+ permeability across the luminal membrane of the cell and increase the shunt conductance for NaCl. According to this interpretation, the decrease in net Na^+ transport would result from a decrease in the intracellular Na^+ pool available for transport. The change in intercellular shunt conductance would offset the change in the apical membrane conductance for electrolytes, so that the measured transepithelial resistance and passive fluxes of Na^+ and Cl^- would remain unchanged, although the pathway would be different. The mechanism of action of dithiothreitol and amphotericin B in this model would be the same as suggested for the previous model. It is of note that this postulate requires a dual effect of heavy metals on membrane transport, the net effect of which would not be readily distinguishable from our first proposal using current technology.

There is no a priori reason to assume that heavy metal-induced alterations in membrane function must result from an interaction between heavy metals and the apical or antiluminal membranes of the cell. It is possible that either uranyl nitrate or $HgCl_2$ may gain entrance to the interior of the cell and, from an intracellular locus, interact with enzyme systems that provide energy for the transport processes per se. To account for the prompt inhibition of SCC after mucosal, but not serosal addition of heavy metals would require a differential rate of heavy metal entry into the cell across the luminal and antiluminal membranes of the cell. Because the serosal side of the epithelial cell of the epithelium chosen for study forms a complex with basement membrane myo-mesothelial elements, it may be presumed to be replete with many more sites capable of complexing heavy metals and, thus, preventing their entry into the cell. Dithiothreitol gains entrance into the cell only from the mucosal side for similar reasons. Amphotericin B reverses the inhibition of Na^+ transport after membrane exposure to uranyl nitrate or $HgCl_2$ by increasing the concentration of intracellular Na^+ available to a partially inhibited transport process. However, there is evidence in other cell types that uranyl nitrate exerts its major effect on cellular transport and cellular respiration only by interacting with the cell surface and does not readily permeate into the cell interior (BARRON et al., 1948; DEMIS et al., 1954; ROTHSTEIN, 1959). Although we have no direct evidence for or against an accumulation of uranyl nitrate within the epithelial cells of the turtle bladder, the lack of an inhibitory effect on SCC after prolonged serosal exposure to uranyl nitrate and the fact that dithiothreitol can rapidly reverse the inhibition of SCC as a consequence of prolonged mucosal exposure to uranyl nitrate, suggest that this heavy metal ion does not accumulate to a significant degree within the epithelial cell. On the other hand, $HgCl_2$ does gain access to the interior of many cell types (WEBB, 1966). In the muscle cells obtained from rat diaphragm, $HgCl_2$ may have both an effect on plasma membranes as well as an effect on cellular enzyme systems after prolonged exposure to this heavy metal (ROTHSTEIN, 1959). The initial and rather rapid accumulation of $HgCl_2$, presumably at superficial membrane sites, results in the inhibition of glucose transport into the cell. This initial phase of $HgCl_2$ accumulation, and its physiologic consequences, can be completely reversed by exposure of the cells to cysteine. A second, slower, component of $HgCl_2$ uptake by these muscle cells eventually leads to an inhibition of oxidative respiration. Neither this latter accumulation of $HgCl_2$, nor its physiologic sequelae can be reversed by exposure to cysteine. In the studies detailed above, performed using the turtle urinary bladder, parallel observations were made. The inhibition of SCC after relatively brief periods of exposure (30-90 min) to $HgCl_2$ was completely reversed by dithiothreitol. In contrast, after prolonged mucosal exposure (120 min, or greater) to $HgCl_2$, the induced alterations in SCC were only partially reversed by dithiothreitol. After serosal exposure of the turtle bladder to $HgCl_2$, the resultant modest inhibition of SCC was also not reversible by treatment with dithiothreitol. These results demonstrate that $HgCl_2$ exerts its effects on ion transport initially at the apical membrane of the epithelial cell, but does eventually gain access into the cell and alters the intracellular enzyme systems, which provide energy for ion transport.

Conclusion. Uranyl nitrate and $HgCl_2$ inhibit Na^+ transport without disrupting the epithelium as a barrier to the passive flow of small ions, and this inhibition is reversible. The most likely site of heavy metal-membrane interaction is within the apical membrane of the epithelial cell and, more than likely, involves complex formation with sulfhydryl groups of macromolecules associated with Na^+ transport. Although the interaction of the heavy metals with these macromolecules has not as yet been characterized, analogies may be drawn from the extant literature. For example, $HgCl_2$ lowers the oxygen affinity of hemoglobin, whereas organic mercurials, such as *para*-chloromercuribenzoate (PCMB), do not alter oxygen affinity (RIGGS, 1952; RIGGS and WALBACH, 1956). This differential effect of organic versus inorganic mercurials may result from the capacity of inorganic mercurials to form cyclic, or polymercaptides, whereas organic mercurials can form only monomercaptides. The spatial arrangement of sulfhydryl groups within the heme molecules are such that bridge formation can readily occur in the presence of inorganic, but not organic mercurials (RIGGS, 1959). This bridge formation may alter the tertiary structure of the reactive site of the heme molecule, resulting in an alteration of its affinity for oxygen in the presence of inorganic mercurials. In toad and turtle urinary bladders, PCMB does not inhibit Na^+ transport (SCHWARTZ, 1975; FRENKEL et al., 1975), in contrast to the demonstrated effects of $HgCl_2$ on Na^+ transport. It is suggested, therefore, that $HgCl_2$, and probably uranyl nitrate, alter the affinity of the Na^+ sites in the apical membrane by forming heavy metal-captide bridges that change the tertiary structure of the reactive site for transport. That this interaction of the heavy metal and the transport site is not an irreversible destruction of the site is indicated by the rapid and complete reversal of the heavy metal-induced alteration in Na^+ transport associated with the addition of dithiothreitol to the bladders.

B. The *in vivo* Preparation

The results obtained with an *in vitro* preparation may be directly
applied to observations made in the intact animal after the administration of heavy metals. Based on the *in vitro* effects of heavy
metals, one would predict the following kinds of abnormalities: (1) a
diminution in net Na^+ reabsorption resulting in an enhanced fractional
excretion of Na^+, and a decline in the ability to concentrate the
urine, since this process requires the active transport of Na^+; and,
(2) a decrease in absolute and fractional fluid absorption by tubule
segments of the nephron in association with alterations in the concentration and/or load of Na^+ at specific anatomic sites along the
nephron.

1. Whole Animal Studies

The acute intravenous administration of uranyl nitrate, 10 mg/kg of
body weight, was characterized in dogs (KLEINMAN et al., 1975). Immediately after the administration of uranyl nitrate, there were
marked increases in both the urine flow rate and the urinary excretion
of Na^+. Urine flow during control periods was 0.54 ± 0.19 ml/min, and
reached a maximal value 3 hours after uranyl nitrate of 1.17 ± 0.10
ml/min. Similarly, the excretion of Na^+ rose from a control value of
28.9 ± 8.6 μmol/min to 137.3 ± 9.0 μmol/min over the same time interval. As indicated in Fig. 11, the calculated fractional excretion of
filtered Na^+ was markedly increased after the administration of uranyl
nitrate. The osmolality of the plasma was not significantly altered,
and, as a result of the progressive decline in urine osmolality, there
was a progressive fall in the urine to plasma osmolality ratio (Fig.
11). Additional studies, carried out for up to 96 hours after the
administration of uranyl nitrate, revealed a continued depression in
urine osmolality. As other parameters of renal function, such as
glomerular filtration rate and renal blood flow, continued to be ab-
normal, the increase in Na^+ excretion and urine flow became less
marked. Indeed, at 72 to 96 hours after uranyl nitrate, urine flow
and Na^+ returned toward control values.

In the normal dog the fractional reabsorption of Na^+ was greater than
99% complete, demonstrating the avid reabsorption of this ion. The
diminished Na^+ reabsorption after the administration of uranyl nitrate
cannot be explained by any alteration in other parameters of renal
function and is consistent with a direct effect of the heavy metal on
renal tubular epithelium. The ability to concentrate urine, as noted
by a high urine osmolality or a high urine to plasma osmolality ratio,
is dependent on the active reabsorption of Na^+ by tubular epithelium
(WIRZ and DIRIX, 1973). Thus, the alterations in osmolality that
occur after uranyl nitrate may be interpreted as the direct consequence of an alteration in Na^+ transport. Similarly, the increase in
urine flow rate is a reflection of a diminution in both Na^+ reabsorption per se as well as an inability to concentrate urine. These results are indicative of a heavy metal-induced tubular membrane dysfunction. Indeed, alterations in solute and solvent movement, manifested by increased urine Na^+ concentration and decreased urine to
plasma osmolality ratios, are sufficiently characteristic as to be

88

Fig. 11. Effect of
uranyl nitrate, 10
mg/kg body weight, i.v.
on inulin clearance,
excreted fraction of
filtered Na^+ (EFF_{Na+}),
and urine to plasma
osmolality ratio (*U/P
Osmolality*) in 8 dogs
over 3-h time interval.
Each bar represents
mean±SEM

Uranyl Nitrate (10 mg/kg, iv) Induced Acute Renal Failure-
Inulin Clearance, EFF_{Na}, U/P Osmolality

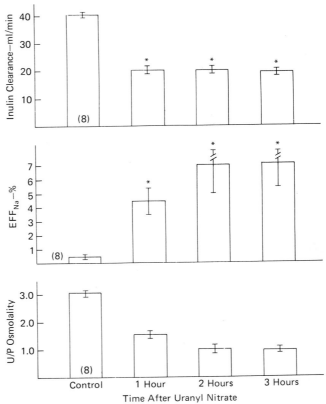

used as criteria in the diagnosis of clinical acute renal failure
(SWANN and MERRILL, 1953; MERRILL, 1960, 1971; PAPPER, 1971; SEVITT,
1959). In addition, parallel abnormalities in fluid and electrolyte
reabsorption have been observed in many models of experimental acute
renal failure (FLAMENBAUM et al., 1972b, 1974; DAUGHARTY et al., 1974).

2. *Micropuncture Studies*

Examinations of alterations in fluid and electrolyte transport using
techniques in which estimates of alterations are obtained based upon
changes in urinary excretion are inherently limited, since they represent the sum total of the function of many thousands of nephron units,
which comprise the total kidney. Therefore, only broad and inferen-
tial conclusions may be based on these results. The technique of *in
vivo* micropuncture of the intact kidney allows further characterization of abnormalities in fluid and electrolyte transport at the single
nephron level. This technique involves the insertion of micropipettes
into individual, anatomically defined nephron segments and, by measuring the concentration of inulin in tubule fluid and plasma and determining the volume of timed collections of tubule fluid, it is possible

to calculate: single nephron glomerular filtration rate (SNGFR),

$$\text{SNGFR} = V \cdot \text{TF}/\text{P}_{\text{IN}},$$

where V = tubule fluid volume flow rate, and $\text{TF}/\text{P}_{\text{IN}}$ = tubule fluid to plasma inulin concentration ratio; fractional fluid reabsorption, $1-\text{P}/\text{TP}_{\text{IN}}(100)$; and, absolute fluid absorption, SNGFR-V. We have applied this renal micropuncture technique to the study of tubular membrane dysfunction after the administration of uranyl nitrate in rats.

Estimates of fluid and electrolyte absorption were obtained early (6 h) and late (48 h) after the injection of uranyl nitrate (10 mg/kg body weight) in rats (FLAMENBAUM et al., 1974). Since the heavy metal-induced alterations in membrane function may not be uniform along the length of the tubule, these estimates were obtained in both early segments (proximal tubule) and late segments (distal tubule) of superficial nephrons. In control animals, the fractional reabsorption of glomerular filtrate to the site of puncture in proximal and distal nephron segments was 45 ± 4% and 81 ± 2%, respectively. Although at 6 h after uranyl nitrate, fractional reabsorption was essentially unchanged, at 48 h, it had significantly decreased to 23 ± 4% and 69 ± 2%, in proximal and distal segments. These results indicate that in the normal animal there is progressive reabsorption of glomerular filtrate along the length of the nephron. Early after the administration of uranyl nitrate, the fractional reabsorption of fluid remains unchanged due to a concomitant decrease in fluid delivery (SNGFR) and the absolute reabsorption of tubular fluid. Later in the course of tubular dysfunction, there is a decrease in fractional fluid absorption due to a greater fall in the absolute rate of fluid absorption than in glomerular filtration. Fig. 12 depicts the calculated changes in absolute fluid absorption to the site of micropuncture in the proximal and distal segments of the nephron at the same time intervals of study after the injection of uranyl nitrate. The significant decrease (31%) in absolute fluid absorption to the site of distal micropuncture and the near normal absorption in the proximal tubule at 6 hours indicates that a decrease in absolute fluid absorption occurred between the two nephron sites. This area of the nephron, which consists of the late proximal tubule and the loop of Henle, has also been demonstrated to be the location where subcellular damage is demonstrable early after uranyl nitrate (RYAN et al., 1973). At later time intervals, the alterations in fluid absorption are more severe and diffuse.

Similar alterations in fluid and electrolyte reabsorption have been demonstrated in $HgCl_2$ (FLANNIGAN and OKEN, 1965; FLAMENBAUM et al., 1971; STEINHAUSEN et al., 1969; BIBER et al., 1968; HENRY et al., 1968), dichromate (BIBER et al., 1968; HENRY et al., 1968), glycerol (OKEN et al., 1966), and methemoglobin (JAENIKE, 1969) induced acute renal failure, as well as after ischemic renal damage (SCHNERMANN et al., 1966). The results of these renal micropuncture studies confirm and extend the observations made based on whole kidney investigations and are consistent with the alterations predicted from *in vitro* experiments. According to the proposed schema of the initiation of acute renal failure, the altered tubular membrane function should result in

ABSOLUTE AND FRACTIONAL FLUID ABSORPTION IN THE
PROXIMAL AND DISTAL TUBULE CONTROL AND 6
HOURS AFTER URANYL NITRATE

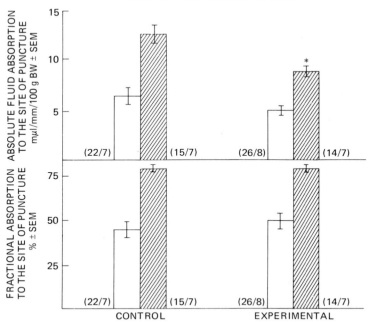

Fig. 12. Absolute and fractional fluid absorption in proximal (*clear bars*) and distal (*crossed-hatched bars*) tubules in control animals and rats 6 h after subcutaneous injection of uranyl nitrate, 10 mg/kg body weight. Numbers within parentheses represent number of punctures performed over number of animals studied. Values are expressed as mean+SEM

a modified tubular fluid milieu at the macula densa. Analysis of tubular fluid obtained from superficial distal tubules 6 h after uranyl nitrate revealed a [Na+] of 116.9 ± 2.5 mEq/1, significantly greater than the mean value of 53.7 ± 1.2 mEq/1 obtained in control rats (Fig. 13). These values are considered to be representative of the [Na+] of tubular fluid at the macula densa of the distal nephron. Thus, evidence is available indicating that alterations in membrane function, both *in vitro* and *in vivo*, due to heavy metals are an integral part of the pattern of acute renal failure.

III. The Renin-Angiotensin System

The most characteristic alterations in renal function, other than changes in fluid and electrolyte transport, are marked decreases in glomerular filtration rate and abnormalities in renal hemodynamics (FLAMENBAUM, 1973). These alterations in glomerular filtration and

Fig. 13. Distal tubule Na⁺ concentration in control rats and
in animals 6 h after subcutaneous injection of uranyl nitrate,
10 mg/kg body weight. Numbers
within bars are number of tubules studied over number of
animals studied. Values are
expressed as mean±SEM

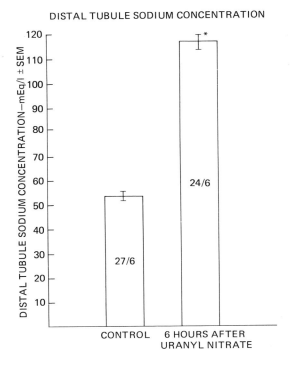

DISTAL TUBULE SODIUM CONCENTRATION

renal blood flow occur as the indirect consequence of altered membrane
function and are mediated by the renin-angiotensin system.

A. General Characteristics

In order to understand the role of the renin-angiotensin system in the
pathogenesis of acute renal failure, a brief background description is
warranted. The reader is referred to recent reviews for further detail (PAGE and McCUBBIN, 1968; LARAGH and SEALEY, 1973). It is important for the purposes of the present discussion, however, to obtain
an overview of the renin-angiotensin system relative to both normal
homeostatic adjustments in renal function as well as in the context of
the pathophysiologic alterations in acute renal failure.

1. Background

The activity of the renin-angiotensin system is intimately related to
the structure of the juxtaglomerular apparatus. In each nephron unit,
the ascending loop of Henle comes into contact with the vascular pole
of the glomerulus; this complex of cells, comprising myoepithelial
cells of the arteriole wall, cells of the distal tubule in contact
with the vascular pole of the glomerulus (the macula densa), and cells
located within the glomerular hilus (lacis cells) constitute the
juxtaglomerular apparatus. The most striking feature of the myoepithelial or juxtaglomerular cells, which are located in the media of

the afferent arteriole, is the abundance of intracellular, secretory granules. The highly developed Golgi apparatus and endoplasmic reticulum, and peripheral myofibrils apparent in these cells, are consistent with their smooth muscle origin and their current endocrine function. The juxtaglomerular cell granule content fluctuates with the secretory rate of renin and appears to be related to the rate of glomerular filtration. The macula densa consists of a group of specialized tubular epithelial cells located in that portion of the nephron in contact with the vascular pole of the glomerulus. Although the general ultrastructural characteristics of these cells are consistent with their being distal tubule cells, there are some fundamental differences. They have fewer mitochondria, the basal infoldings of the plasma membrane are less well developed, the Golgi apparatus is basally located, the granular endoplasmic reticulum is poorly developed, and ribosomes are few as compared to the other distal tubular epithelial cells. These structural differences are not consistent with a reabsorptive function for the macula densa cells of the distal nephron. The lacis cells are located in a triangle formed by the afferent and efferent glomerular arterioles and the macula densa segment of the distal tubule. These cells have a paucity of cytoplasmic organelles, but contain actinomysin-fibrillar structures. At the periphery of the lacis cells there are numerous cytoplasmic projections, and the cells are interdigitated by a basement membrane that is continuous with the membrane lining the visceral epithelial lining of the glomerulus.

The juxtaglomerular apparatus also receives nonmyelinated nerve fibers. These nerve fibers have been demonstrated in the walls of the afferent and efferent arterioles, although the interrelationship of these adrenergic nerve fibers and the arteriole may differ according to the location of the glomerulus within the renal cortex (LJUNGQVIST and WAGERMARK, 1970).

The various components required for the enzymatic generation of angiotensin can be found within the kidney. Using a variety of experimental techniques, renin has been found almost exclusively within the juxtaglomerular apparatus, predominantly within the myoepithelial cells of the afferent arterioles. The secretory granules observed within the myoepithelial cells are renin-containing secretory granules (COOK, 1971) and the myoepithelial cells store as well as secrete renin. Renin is the enzyme that catalyzes the hydrolysis of a decapeptide, angiotensin I, from renin substrate, an α_2-globulin synthesized by the liver. A second enzyme, converting enzyme, which is a carboxypeptidase, is required for the ultimate generation of angiotensin II, an octapeptide. The intrarenal enzymatic generation of angiotensin II necessitates that renin substrate, and converting enzyme, in addition to renin, be present within the kidney. Evidence has been obtained for the presence of converting enzyme, in concentrations high enough to convert angiotensin I formed within the kidney to angiotensin II, in the juxtaglomerular apparatus (GRANGER et al., 1972). Renin substrate must penetrate into the renal interstitium from the renal vascular space, since this substance has been found in significant quantities within the renal lymph (LEVER and PEART, 1972) and within juxtaglomerular cells (SUTHERLAND, 1970). Confirming evidence for the local intrarenal, myoepithelial cell generation of angiotensin has been obtained in support of these speculations (OSBORNE et al., 1975; MENDELSOHN and JOHNSTON, 1972; FINKIELMAN et al., 1972). Furthermore, the enzyme metabolizing angiotensin, angiotensinase, is also present within the renal parenchyma and the juxtaglomerular apparatus (GRANGER et al., 1972; J.J. BROWN et al., 1972).

2. *Physiologic Alterations*

Many studies of the physiologic alteration in renal renin activity have involved determinations of the renal renin content of the entire kidney or of the partially dissected renal cortex. In view of the localization of renin to specific sites within the juxtaglomerular apparatus, it is apparent that such studies may be affected by the presence of nonrenin containing renal tissue. Furthermore, there is evidence in many animal species that renin is not homogeneously distributed through the renal cortex, suggesting that relevant data may be obscured by determining the renin content even of large areas of the renal cortex. In order to more clearly define the interrelationships between alterations in renin activity on the single nephron level and physiologic perturbations, a technique has been developed that allows the quantification of renin within single juxtaglomerular apparatus (DAHLHEIM et al., 1970; GRANGER et al., 1972; FLAMENBAUM and HAMBURGER, 1974). Using renin-renin substrate kinetics, the content of single juxtaglomerular apparatus has been measured as follows (Fig. 14): the animal is anesthetized and a polyethylene cannula inserted into the abdominal aorta below the level of the renal arteries; after 30 minutes, a bolus of silicone-rubber compound is injected, the kidneys extirpated, and quick frozen in a dry-ice acetone mixture; single juxtaglomerular apparatus are microdissected using the colored silicone rubber compound present within the glomerular capillaries as an indication of anatomic boundaries; the single juxtaglomerular apparatus, which may be obtained from superficial or deep renal cortical areas, are placed into 0.2 ml of renin-free, angiotensinase-free renin substrate and sonicated to achieve dispersal of the renin within the substrate; and, the single juxtaglomerular apparatus-renin substrate mixture is then prepared for assay by the quantification of generated angiotensin. The assay may be performed either by using a bioassay technique, as outlined in Fig. 14, or by a radioimmunoassay for generated angiotensin I (FLAMENBAUM and HAMBURGER, 1974). The bioassay and radioimmunoassay techniques give comparable results and allow the quantification of the renin activity in single juxtaglomerular apparatus as nanograms of generated angiotensin per single juxtaglomerular apparatus per hour of incubation (ng/JGA/h).

a. *Renin Activity and NaCl Intake*

Using the technique of renin determinations within single juxtaglomerular apparatus, studies of the role of intrarenal renin in a variety of physiologic and pathophysiologic states may be undertaken. It is well known, for example, that both circulating, peripheral renin activity and renal renin content vary inversely with NaCl intake. To further delineate the effect of variations in NaCl intake on both the quantity of single juxtaglomerular renin activity as well as any

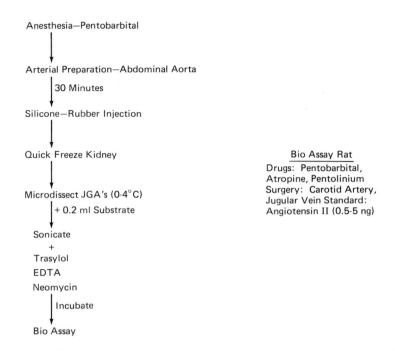

Fig. 14. Schematic outline of determination of renin activity in single juxtaglomerular apparatus (*JGAs*). Bioassay procedure in rat is depicted in right-hand portion of figure

alterations in the intrarenal distribution of renin content, studies were performed in rats maintained on various NaCl intakes. The regular, high, and low NaCl diets contained 287 mEq Na⁺/kg + tap water, 287 mEq Na⁺/kg + 1% saline solution and 2 mEq Na⁺/kg + tap water, respectively. The results of the determination of superficial and deep cortical juxtaglomerular apparatus renin activities in rats maintained on these various Na⁺ diets are presented in Fig. 15. In rats maintained on a regular NaCl diet, the mean superficial juxtaglomerular renin activity was 6.79 ± 0.46 ng/JGA/h, significantly greater than the value of 2.67 ± 0.19 ng/JGA/h obtained from deep cortical juxtaglomerular apparatus, indicating that an intracortical renin gradient is present within the normal rat. As is apparent in Fig. 15, NaCl loading resulted in decreased juxtaglomerular renin activities, and NaCl deprivation resulted in increased juxtaglomerular renin activities. These results are consistent with the well-known inverse relationship between sodium intake and renal renin content. More importantly, significant changes were noted in the relative proportions of superficial and deep juxtaglomerular apparatus renin activities. The quantitative increase in renin activity was more marked in the deep than in the superficial juxtaglomerular apparatus after NaCl deprivation, and the quantitative decrease in renin activity was more marked in the superficial juxtaglomerular apparatus after NaCl loading, as indicated by the relative ratios of superficial to deep renin activities. The mean ratio after high NaCl diet was 1.00 ± 0.07,

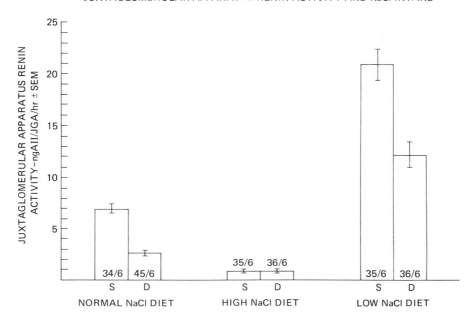

JUXTAGLOMERULAR APPARATUS RENIN ACTIVITY AND NaCl INTAKE

Fig. 15. Superficial (S) and deep (D) juxtaglomerular apparatus renin activity as function of NaCl intake. Numbers within bars are numbers of juxtaglomerular apparatus studied over number of rats studied in each group. Values are expressed as mean±SEM

significantly different from the ratios observed in rats on a low NaCl diet, 1.75 ± 0.12, and regular NaCl diet, 2.52 ± 0.09. Thus, there are clear differences between the response of superficial and deep nephrons in both their net renin activity and their response to variations in NaCl intake.

According to the proposed schema for the initiation of acute renal failure, there must be a demonstrable interrelationship between juxtaglomerular renin activity and both renal hemodynamics/glomerular filtration as well as alterations in tubular epithelial reabsorption of fluid and electrolyte. That the observed changes in both renin content and its intrarenal distribution are consistent with other alterations in parameters of renal function is apparent from investigations of single nephron function under conditions of varying NaCl intake. Certain inferences of physiologic import concerning animals on various NaCl diets may be based on the following assumptions:

1. Superficial single nephron glomerular filtration rate is plasma flow-dependent (BRENNER et al., 1972), such that alterations in glomerular plasma flow are directly reflected by changes in superficial single nephron glomerular filtration rate.

2. The generation of angiotensin on the local nephron level primarily affects afferent arteriolar vasoconstriction in the superficial nephron units, which, in addition to diminishing glomerular plasma flow, alters tubular fluid absorption by changes in intratubular hydrostatic pressure and peritubular oncotic pressure (BRENNER et al., 1969).
3. An inverse relationship exists, therefore, between superficial single nephron glomerular filtration rate and juxtaglomerular apparatus renin activity, since single nephron glomerular filtration rate is plasma flow-dependent and angiotensin decreases glomerular plasma flow.
4. Techniques are sufficiently comparable to assume that juxtaglomerular apparatus identified as being superficial or deep in location are representative of the groups of nephron units identified as superficial or juxtamedullary by either direct or indirect studies of single nephron glomerular filtration rate.

Although surface nephrons are plasma flow-dependent, current evidence suggests that nephrons located within the deep cortex (BARRATT et al., 1973; BRUNS et al., 1974) are not plasma flow-dependent. Morphologic differences between superficial cortical and juxtamedullar arterioles have been described (LJUNGQVIST and WAGERMARK, 1970; BEEUWKES, 1971; LJUNGQVIST, 1975; BEEUWKES and BONVENTRE, 1975): superficial cortical arteriole-glomerular units are characterized by an afferent arteriole, glomerular capillaries, and efferent arteriole occurring in series, with the efferent arteriole diameter consistently less than the afferent arteriole diameter; juxtamedullary arteriole-glomerular units were characterized by glomerular capillaries in parallel, with the potential for a continuous afferent-efferent segment, which may be under the control of the lacis cells in the glomerular hilum, and the ef efferent-afferent vessels are usually of equal diameter. Since single nephron glomerular filtration rate in deep nephron units may not be directly glomerular plasma flow-dependent, and in view of the other noted differences between superficial and deep nephron units, juxtaglomerular apparatus renin activity-associated alterations in single nephron glomerular filtration rate within juxtamedullary nephrons may be other than those predicted for superficial nephron units. Some of these differences are depicted diagramatically in Fig. 16.

Based on the foregoing assumptions, an inverse relationship should exist between single nephron glomerular filtration rate and juxtaglomerular apparatus renin activity. As shown in Table 2, using various techniques, superficial single nephron glomerular filtration rate is normally less than juxtamedullary single nephron glomerular filtration rate in rats on a "normal" NaCl diet. Indeed, the ratio of juxtamedullary-superficial single nephron glomerular filtration rate approximates the superficial-deep ratio of juxtaglomerular apparatus renin activity, and such a similarity demonstrates that there is an inverse relationship between filtration rate and renin activity. The physical factors responsible for the changes in filtration rate relative to alterations in renin activity in response to a high NaCl diet are depicted in Fig. 16. Single nephron glomerular filtration rate (SNGFR) is determined by the net ultrafiltration pressure (P_{UF}) and the ultrafiltration coefficient (K_f), which is the product of the

SUPERFICIAL

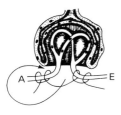

DEEP

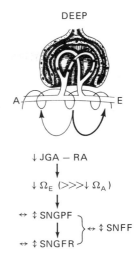

A ←→ E

$\downarrow$ JGA−RA

$\downarrow$ Ω_A

$\uparrow$ SNGPF
$\left.\phantom{\begin{matrix}a\\b\\c\end{matrix}}\right\}$ $\leftrightarrow$ SNFF
$\uparrow$ SNGFR

$\downarrow$ JGA − RA

$\downarrow$ Ω_E ($>>>\downarrow$ Ω_A)

$\leftrightarrow$ $\updownarrow$ SNGPF
$\left.\phantom{\begin{matrix}a\\b\\c\end{matrix}}\right\}$ $\leftrightarrow$ $\updownarrow$ SNFF
$\leftrightarrow$ $\updownarrow$ SNGFR

$$SNGFR = K_f \cdot <P_{UF}>$$

$$K_f = k \cdot S \ (S \ maybe \ \alpha \ to <P_{GC}>)$$

$$<P_{UF}> = (P_{GC} - P_T) - (\Pi_{GC} - \Pi_T)$$

$\textcircled{1}$ $\Pi_T \cong 0$

$\textcircled{2}$ $\Omega_A = \dfrac{<AP> - <P_{GC}>}{GBF}$

$\Omega_E = \dfrac{<P_{GC}> - P_C}{EABF}$

$\textcircled{3}$ $\downarrow \Omega_A \rightarrow \uparrow GBF, \uparrow <P_{GC}>, \uparrow <P_{UF}>, \uparrow SNGFR$
  (dependent on ΔP_T, Π_{GC})

$\textcircled{4}$ $\downarrow \Omega_E \rightarrow \uparrow EABF, \downarrow <P_{GC}>, \downarrow <P_{UF}>, \downarrow SNGFR$
  (dependent on ΔP_T, Π_{GC})

Fig. 16. Interrelationship of high NaCl diet, superficial or deep juxta-glomerular apparatus renin activity (*JGA-RA*), and single nephron function. Abbreviations: A= afferent arteriole; E= efferent arteriole; SNGFR= single nephron glomerular filtration rate; SNGPF= single nephron glomerular plasma flow; and, SNFF= single nephron filtration fraction (*SNGFR/SNGPF*). See text for other abbreviations and complete description of figure

Table 2. Superficial (S) and juxtamedullary (JM) single nephron glomerular filtration rates in normal rats as determined by direct micropuncture techniques

| References | Single nephron filtration rate (nl/min/g kidney weight) | | 1/JM:S |
	S	JM	
JAMISON, 1970	25.6	60.2	2.35
JAMISON and LACY, 1971	21.3	58.3	2.74
STUMPE et al., 1969	30.5	59.7	1.96

glomerular capillary surface area available for filtration (S) and
the hydraulic conductivity of the composite glomerular membrane (k).
P_{UF} is determined by the difference between the net hydrostatic
forces (glomerular capillar hydrostatic pressure, P_{GC}, minus intratubular hydrostatic pressure, P_T) and the net oncotic forces (glomerular capillary oncotic pressure, π_{GC}, minus tubular oncotic pressure,
π_T, which is vanishingly low, since filtrate is protein free). Afferent arteriolar resistance (Ω_A) is the difference between systemic
arterial pressure (AP) minus P_{GC} divided by glomerular blood flow
(GBF); efferent arteriolar resistance (Ω_E) is equal to P_{GC} minus peritubular capillary pressure (P_C) divided by efferent arteriolar blood
flow (EABF). The consequences of the changes in afferent or efferent
arteriolar resistance due to high NaCl diet-induced decreased renin
activity are depicted in Fig. 16. It would be predicted, that while
the predominant renin-related change in superficial units would be a
decrease in afferent arteriolar resistance, the predicted renin-
related change in deep nephron units would be a decrease in efferent
arteriolar resistance, which, depending upon whether blood flowed
through the afferent-efferent arteriole shunt or through the glomerular capillaries, could result in no change, increase, or decrease in
filtration rate.

Consistent with these predictions have been the findings in most
studies that superficial single nephron filtration rates increased
after NaCl loading, whereas the changes in juxtamedullary single nephron glomerular filtration rate varied among the studies. The factors
responsible for the variability in the response of deep nephron units
to NaCl loading are:

1. The basal distribution of juxtaglomerular apparatus renin activity
   and NaCl homeostasis,
2. Age of the rat at the time of study (BAINES, 1973),
3. The degree of reorientation of blood flow as the consequence of
   decreased superficial nephron unit renin activity resulting in a
   preferential increase in outer cortical blood flow ("superficial
   steal"),
4. A differential arteriolar responsiveness between the superficial
   and deep nephron units to the vasoconstrictor effects of angiotensin.

There are no studies of nephron filtration rate in rats maintained on
severe NaCl restriction. Using the present assumption, and observed
changes in juxtaglomerular apparatus renin activity, the predicted
alterations would be the converse of those indicated in Fig. 16.
Decreases in superficial single nephron glomerular filtration rate
would occur in rats on a low NaCl diet, tending to limit NaCl filtration and excretion. The relationship between juxtamedullary single
nephron glomerular filtration rate and deep juxtaglomerular apparatus
renin activity is less clear. If, as predicted by GAVRAS et al.
(1970), the predominant effect of angiotensin is on the afferent arterioles of superficial nephrons and on the efferent arteriole of juxtamedullary nephrons, then, under conditions of NaCl restriction, there
would be relative preservation of juxtamedullary single nephron filtration rate at the expense of decreases in superficial nephron

filtration rate. This would allow preferential maintenance of sodium filtration in the deep cortical nephrons, with long loops of Henle, and thereby conserve NaCl.

b. *Renin Activity and Carotid Manipulation*

That factors other than NaCl may affect renin activity, and thereby alter renal hemodynamics, is suggested by the results obtained after measurements of juxtaglomerular apparatus renin activity following acute, single carotid artery occlusion. When the carotid artery, as opposed to the abdominal aorta (Fig. 14), was cannulated, the result-ing mean superficial and deep juxtaglomerular renin activities were 7.22 ± 0.40 and 6.27 ± 0.13 ng/JGA/h, respectively. The increase in deep renin content in these animals resulted in a decrease in super-ficial to deep renin gradient to 1.18 ± 0.08 (control = 2.52 ± 0.09). In contrast, values not significantly different from control were observed if a chronic carotid artery cannula, placed 3 to 5 days pre-viously, was used for injection of silicone-rubber compound (FLAMENBAUM and HAMBURGER, 1974). Because of the possible introduction of arti-fact into the parameters under study, carotid manipulation is avoided by many investigators. Measurements of the intrarenal distribution of blood flow before and after acute carotid artery manipulation de-monstrate that with carotid manipulation there is a redistribution of blood flow from outer to inner cortical regions (LOGAN et al., 1971). On the other hand, alterations in the intrarenal distribution of blood flow may be avoided by using a chronically implanted carotid artery cannula for these studies (WALLAN et al., 1971). Acute manipulation of the carotid artery may, therefore, alter the intrarenal distribu-tion of both blood flow and renin activity.

Figure 17 diagrammatically depicts the mechanism by which acute carotid artery manipulation may result in alteration in both renal hemodynamics and renin activity. An inverse relationship has been demonstrated between carotid sinus pressure and renal nerve sympathetic activity, mediated by the carotid baroreceptor mechanism (CLEMENT et al., 1972; NINOMIYA et al., 1971; NINOMIYA and IRISAWA, 1969; KEZDI and GELLER, 1968; GILMORE, 1964). In addition, increased renal sympathetic nerve activity results in increased renin-angiotensin system activity (MOGIL et al., 1969; DAVIS, 1971; ZEHR and FEIGL, 1973), which is independent of changes in glomerular filtration (JOHNSON et al., 1971). Increased renal nerve activity results in increased renal vascular resistance (BUNAG et al., 1966; POMERANZ et al., 1968), which is quantitatively more important at systemic blood pressures greater than 100 mmHg (LAGRANGE et al., 1973). Carotid manipulation may, therefore, as the result of increased sympathetic nerve tone to the nephron, increase renin synthesis and release, resulting in a decline in outer cortical blood flow. In addition, a direct decrease in outer cortical blood flow due to increased renal nerve activity may increase renin release. The net result of either, or both, pathways would be no net change in juxtaglomerular apparatus renin activity, since net content would be a reflection of both increased synthesis and release. The effects of the increased sympathetic tone on the relative vascular resistance across juxtamedullary nephron units, as compared to superficial units (Fig. 16), would be less marked, resulting in a preferential shift in

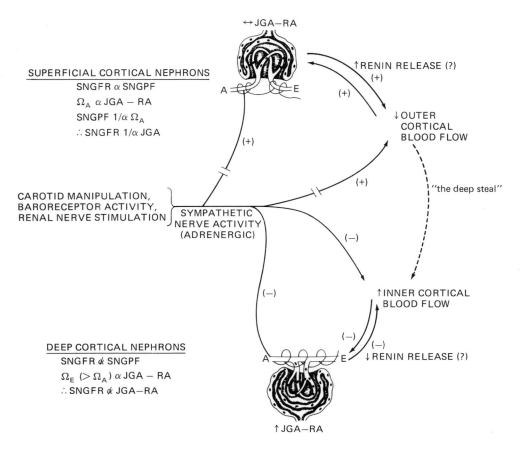

$\leftrightarrow$ JGA—RA

SUPERFICIAL CORTICAL NEPHRONS

SNGFR α SNGPF

$\Omega_A \; \alpha$ JGA — RA

SNGPF $1/\alpha \; \Omega_A$

$\therefore$ SNGFR $1/\alpha$ JGA

$\uparrow$ RENIN RELEASE (?)

(+)

$\downarrow$ OUTER CORTICAL BLOOD FLOW

(+)

(+)

(+)

"the deep steal"

CAROTID MANIPULATION, BARORECEPTOR ACTIVITY, RENAL NERVE STIMULATION

SYMPATHETIC NERVE ACTIVITY (ADRENERGIC)

(−)

(−)

(−)

$\uparrow$ INNER CORTICAL BLOOD FLOW

(−)

(−)

$\downarrow$ RENIN RELEASE (?)

DEEP CORTICAL NEPHRONS

SNGFR α SNGPF

$\Omega_E \; (> \Omega_A) \; \alpha$ JGA — RA

$\therefore$ SNGFR α JGA—RA

$\uparrow$ JGA—RA

Fig. 17. Diagram of proposed mechanism by which carotid artery manipulation may result in alterations in both juxtaglomerular apparatus renin activity (*JGA-RA*) and renal hemodynamics. See Fig. 16 and text for explanation of abbreviations and complete description of figure

blood flow from outer to inner cortical regions ("deep steal"). This net increase in inner cortical blood flow to these deep cortical nephrons would tend to diminish renin release and, therefore, increase deep juxtaglomerular apparatus renin activity.

B. The Renin Angiotensin System in Acute Renal Failure

That the renin angiotensin system mediates the pathophysiologic manifestations of acute renal failure was suggested by GOORMAGHTIGH more than 25 years ago (GOORMAGHTIGH, 1940, 1945, 1947). In reviewing the renal histopathology of patients dying from acute renal failure, he observed glomerular ischemia, juxtaglomerular hypertrophy, and increased juxtaglomerular cell granularity. Based on these observations, he concluded that, "the enhanced endocrine vasopressive action of the renal arterioles seems to be the determining factor in producing the ischemic kidney" (GOORMAGHTIGH, 1945) and that, "this vasopressive substance [*was*] liberated in excess and [*caused*] a persistent spasm at the vascular pole of the glomerular tufts." He further suggested that, "this substance acts on the neighboring smooth muscle cells [*of the arterioles*]" and "overflows in the general circulation" (GOORMAGHTIGH, 1947). The relationship of the anatomic structure of the juxtaglomer-ular apparatus with the "musculo-endocrine" mechanism, which he was proposing, led him to conclude that "the glomeruli and the tubules must not be considered independently," since "the existence of a con-nection at the level of the macula densa suggests a possible func-tional interrelationship between the two parts of the nephron."

1. *Circulating, Plasma Renin Activity in Acute Renal Failure*

Further support for a role of the renin-angiotensin system in the pathophysiology of acute renal failure has been obtained from the de-monstration of increased activity of this system in both clinical and experimental acute renal failure. Several groups have observed in-creased peripheral levels of renin or angiotensin in patients with acute renal failure associated with a wide variety of initiating events (TU, 1965; MASSANI et al., 1966; BROWN et al., 1970; OCHOA et al., 1970). The general pattern of results indicated that the activ-ity of the renin-angiotensin system, as estimated from circulating levels of renin or angiotensin, was most marked during the oliguric phase of acute renal failure and then returned toward normal after the onset of the recovery phase. Furthermore, a general, inverse correl-ation was inferred between the levels of activity of the renin-angiotensin system and the prognosis for recovery from acute renal failure by BROWN and co-workers (1970), since patients dying from acute renal failure had a higher level of circulating renin levels than patients recovering from acute renal failure. Increased renin-angiotensin system activity has also been observed using estimates of plasma renin activity in experimental acute renal failure after the administration of methemoglobin (RUIZ-GUINAZU, 1971), glycerol (DiBONA and SAWIN, 1971; W.C.B. BROWN et al., 1972; RAUGH et al., 1973), uranyl nitrate (FLAMENBAUM et al., 1972b, 1975), and $HgCl_2$ (MATHEWS et al., 1974).

Studies of intrarenal renin content and circulating renin activities have suggested that the renin-angiotensin system participates in the pathophysiology of acute renal failure via intrarenal renin, and that increased circulating levels reflect an "overflow phenomenon," as suggested by GOORMAGHTIGH and others (FLAMENBAUM et al., 1972a; FLAMENBAUM, 1973). Marked suppression of both intrarenal and circu-lating renin activities by NaCl loading (FLAMENBAUM et al., 1972a) diminishes the alterations in renal function after the induction of experimental acute renal failure (HENRY et al., 1968; RYAN et al., 1973; DiBONA et al., 1971; McDONALD et al., 1969; THIEL et al., 1970). In contrast, the acute administration of saline and desoxycorticosterone suppressed only plasma renin activity and not intrarenal renin activ-ity (FLAMENBAUM et al., 1973), and did not significantly alter the course of glycerol-induced myohemoglobinuric acute renal failure. This observation suggested that it was renal renin, rather than

circulating renin activity, that participates in the pathophysiology
of acute renal failure. Partial suppression of renal renin, resulting
from the chronic administration of an increased KCl intake, resulted
in only partial protection against the development of acute renal
failure (FLAMENBAUM et al., 1973), and immunization against heterolog-
ous renin suppressed plasma, but not intrarenal renin and did not
confer protection (FLAMENBAUM et al., 1972a). In view of these results, it may be concluded that increases in the peripheral, circulating plasma renin levels reflect an "overflow" into the general
circulation of renin released from juxtaglomerular apparatus. It is
not surprising, in this regard, that evaluations of the role of the
renin-angiotensin system in acute renal failure, based solely on data
derived from measurements of circulating renin, may fail to indicate
a role for increased renin activity (MATHEWS et al., 1974; POWELL-
JACKSON et al., 1972; OKEN et al., 1975a,b).

2. *Intrarenal Renin Activity in Acute Renal Failure*

Since estimates of renal renin activity, based on assay of the entire
kidney or renal cortex, may be influenced by the presence of nonrenin
containing tissues, we have studied the role of intrarenal renin in
acute renal failure by analyzing the alterations in juxtaglomerular
apparatus renin activity. Rats were injected with uranyl nitrate
(10 mg/kg body weight, subcutaneously) and single juxtaglomerular apparatus harvested as indicated above 2, 4, 6, 24, and 48 h after the
induction of acute renal failure. Plasma renin activity was also
determined in separate groups of rats 6, 24, and 48 h after the administration of uranyl nitrate. The results of this study are depicted
in Fig. 18. Promptly after the administration of uranyl nitrate,
there were marked increases in juxtaglomerular apparatus renin activity as determined in both the superficial and deep portions of the
renal cortex. These elevations were maintained for up to 6 h after
the induction of acute renal failure. At later time intervals, 24
and 48 h after the injection of uranyl nitrate, juxtaglomerular apparatus renin activities returned to values that were not significantly
different from control. In contrast, plasma renin activity that had
doubled by 6 h after uranyl nitrate, continued to increase at 24 and
48 h after the induction of acute renal failure. The marked increases
in both circulating and intrarenal renin activities during the early
phase of uranyl nitrate-induced acute renal failure is consistent with
a role for increased renin-angiotensin system activity in the initiation of the uranyl nitrate model. The subsequent course of changes
in juxtaglomerular vis-à-vis circulating renin activities is the
subject of some speculation. This discrepancy may be the result of an
artifactual selection of juxtaglomerular apparatus with normal renin
activity. The technique of microdissection of juxtaglomerular apparatus requires the filling of the vascular space of glomeruli with a
silicone-rubber compound for ease of identification. In view of the
marked abnormalities in renal hemodynamics, which characterizes acute
renal failure (see below), nephron units with the least severe vasoconstriction may have been preselected for study at the later time
intervals. This would exclude the most involved nephron units from
study, and would preselect those nephron units that were either least
involved or had entered into a recovery phase. It has been estimated

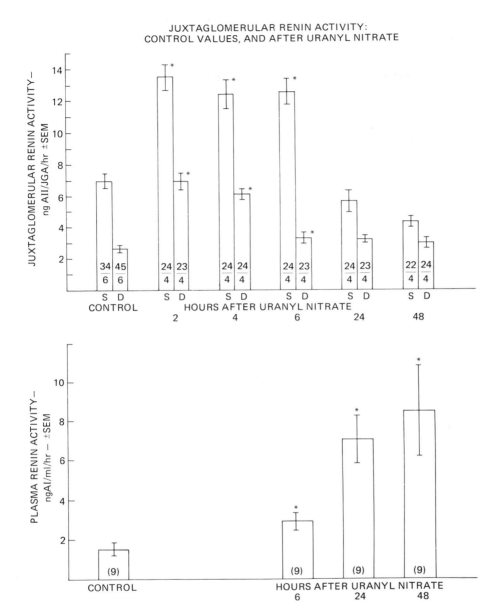

Fig. 18. Alterations in superficial (S) and deep (D) juxtaglomerular apparatus and plasma renin activities after uranyl nitrate, 10 mg/kg body
weight, subcutaneously. Numbers within bars represent number of determinations over number of rats studied, or number of animals studied. Renin
activities are expressed as the generation of angiotensis I (A I) or II
(A II)/h of incubation in 1 ml of plasma or in a single juxtaglomerular
apparatus (JGA). Values are expressed as mean+SEM

that, on an equal volume basis, the ratio of renin in a juxtaglomerular apparatus to that within an equal volume of blood is approximately 10^{16} (THURAU and MASON, 1974). The maintained increase in plasma renin activity, despite the lack of an elevation in juxtaglomerular apparatus renin activity, may thus represent the overflow from a limited number of involved units because of the disparity in renin content.

Alternatively, the increase in plasma, but not intrarenal renin activity, at 24 and 48 h after uranyl nitrate may be a reflection of the differential roles for these two sites of renin activity in the later phase of acute renal failure. Thus, the decline in juxtaglomerular renin activity may be viewed as a lack of a role for intrarenal renin in the maintenance phase of acute renal failure, and the persistence of an elevated plasma renin activity ascribed to a role for angiotensin in systemic vascular tone.

It has been demonstrated that changes in renal renin on the nephron level, in the juxtaglomerular apparatus, may alter single nephron function (SCHNERMANN et al., 1966, 1970, 1971, 1973; THURAU and SCHNERMANN, 1965; THURAU, 1966, 1972a,b; CORTNEY et al., 1966; THURAU et al., 1972; NAVAR et al., 1974; WRIGHT and SCHNERMANN, 1974). It has been suggested that a mechanism, "tubuloglomerular feedback," accounts for the interrelationship observed between nephron function and renin-angiotensin system activity. It has been proposed that alterations in tubular fluid entering the macula densa segment of the distal nephron are associated with increases in juxtaglomerular apparatus renin activity and decreased single nephron glomerular filtration rate. To study this, the composition of fluid entering the macula densa segment of the distal nephron may be altered in a controlled fashion using a micropuncture/microperfusion technique. It has been demonstrated that an increase in tubular fluid sodium chloride delivery/concentration at and transport across macula densa cells is the stimulus responsible for feedback. As a result of this stimulus sensed at the macula densa, there is an increase in renin release from the juxtaglomerular apparatus, the local generation of angiotensin, increased glomerular arteriolar tone, diminished glomerular plasma flow rate, and decreased glomerular filtration rate (DEEN et al., 1973). Similarly, as noted above, an inverse correlation between filtration rate and renin activity on the single nephron level has been observed in states of altered sodium intake. It is also apparent that the sensitivity ("tonic state") of this response is inversely related to the sodium chloride balance of the animal (DEV et al., 1974; KAUFMAN, HAMBURGER, and FLAMENBAUM, unpublished observations). This adjustment of glomerular function (single nephron filtration rate) to tubular function (net sodium chloride or solute and solvent movement) by an endocrine mechanism (the renin-angiotensin system) affecting muscle cells (glomerular arterioles) parallels the predictions made by GOORMAGHTIGH on purely anatomic grounds.

It has been suggested previously that the pathophysiologic manifestations of acute renal failure are a representation of a normal, albeit markedly increased, activity of the tubuloglomerular feedback mechanism (THURAU, 1970; SCHNERMANN et al., 1966; FLAMENBAUM et al., in press). A greater than twofold increase in sodium chloride concentration within tubular fluid in the distal nephron, representative of changes in tubular fluid concentration within the macula densa segment of the nephron, in both uranyl nitrate and ischemic acute renal failure, as well as an inverse correlation between distal tubular fluid sodium concentration and glomerular filtration rate has been observed (SCHNERMANN et al., 1966). Furthermore, it has been demonstrated that this mechanism is intact in acute renal failure (Fig. 19). According to the proposed operation of the feedback mechanism, increases in sodium concentration resulting from retrograde perfusion of the macula densa segment of the distal nephron with 150 mM Ringer's solution should decrease single nephron filtration rate (as estimated by tubular fluid flow rate from early proximal tubule segments). Conversely,

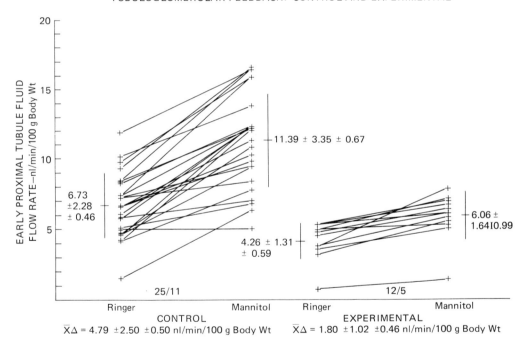

Fig. 19. Demonstration of tubuloglomerular feedback in control rats and rats receiving uranyl nitrate, 10 mg/kg body weight, subcutaneously 6h previously (*experimental*). Using multiple micropipette technique, macula densa segment of distal nephron was perfused in retrograde manner with either Ringer's solution containing 150 mM Na[+] or with isotonic mannitol solution. Proximal tubule fluid flow rate in earliest puncturable superficial segment was taken as estimate of glomerular filtration rate and is expressed as mean+SD+SEM. Average change in proximal tubule fluid flow rate is indicated beneath each study group, and number of punctures over number of animals studied is also indicated. In both control and experimental groups, sodium-free mannitol perfusion resulted in increase in tubular fluid flow rate. Adapted from TAKABATAKE et al. (1974)

decreasing the sodium signal at the macula densa by retrograde perfusion with sodium-free isotonic mannitol solution should increase
proximal fluid flow rate. As may be seen in Fig. 19, these predicted
alterations in filtration rate occurred in both normal, control rats
as well as in animals that had received uranyl nitrate 4 to 6 hours
previously. These results are consistent with an intact functioning
of the tubuloglomerular feedback mechanism in acute renal failure.
It is apparent that changes in tubular epithelial function resulting
in altered tubular fluid composition may be the initial event activating the renin-angiotensin system, via the mechanism of tubuloglomerular feedback in view of the altered tubular function demonstrated in
many models of acute renal failure (BIBER et al., 1968; FLAMENBAUM et
al., 1972, 1974; DAUGHARTY et al., 1974; HENRY et al., 1968; BARENBERG
et al., 1968). In addition, it has been demonstrated that marked increases in juxtaglomerular apparatus renin activity, analogous to
those observed in acute renal failure, may be induced by retrograde
microperfusion of distal nephron segments with fluid having a high
sodium chloride concentration (THURAU and MASON, 1974). Thus, the
altered renal hemodynamics and decreased glomerular filtration rate,
with the consequent azotemia, may be viewed as a "normal" response to
an alteration in tubular function induced by nephrotoxins or ischemia
during the initiating phase of acute renal failure.

IV. Renal Hemodynamics and Acute Renal Failure

The primacy of alterations in renal hemodynamics as the mechanism
responsible for the very common finding of diminished effective glomerular filtration in acute renal failure has been well established
(BULL et al., 1950). There is prompt and continued decrease in total
renal blood flow during the initiating and maintenance phases, with
return toward normal during the recovery phase, of acute renal failure.
Recent studies of renal hemodynamics have avoided estimates of renal
blood flow using standard clearance techniques. The development of
alternative methods, such as inert gas washout, indicator dilution,
electromagnetic flowmeter, and radiomicrospheres, for determination of
renal hemodynamics have allowed further characterization of both total
renal blood flow and intrarenal distribution of blood flow in acute
renal failure (AYER et al., 1971; FLAMENBAUM et al., 1972b, 1974;
KLEINMAN et al., 1975; REUBI et al., 1962; THORBURN et al., 1963;
HOLLENBERG et al., 1968, 1970).

A. Clinical Acute Renal Failure

Marked depressions in total renal blood flow have been observed by
HOLLENBERG and co-workers (1968, 1970) and LADEFOGED (LADEFOGED and
WINKLER, 1970; PEDERSON and LADEFOGED, 1973) in clinical acute renal
failure. These investigators also demonstrated that there was a
marked and sustained reduction in cortical perfusion, suggesting that
in addition to the decrease in total renal blood flow, there also
occurred a redistribution of intrarenal blood flow away from outer

cortical regions. Further support for this redistribution was evident
in comparing the reductions in total blood flow of patients with acute
and chronic renal failure. While total renal blood flow was reduced
to a comparable degree in patients with acute and chronic renal failure, there were marked decreases in cortical perfusion in patients
with acute renal failure with unmeasurable rates of glomerular filtration. In contrast, in patients with chronic renal failure, and measurable glomerular filtration, outer cortical perfusion was maintained,
albeit diminished (HOLLENBERG et al., 1968, 1970). In addition, the
renal hemodynamic abnormalities in patients with acute renal failure
secondary to diverse initiating events were strikingly similar, suggesting that a vasomotor mechanism is the "final common pathway" for
the pathophysiology of acute renal failure. The association of the
return of total renal blood flow to normal and the reappearance or
increase of cortical perfusion with the return of glomerular filtration are consistent with a primary role for altered renal hemodynamics
in the recovery from acute renal failure (HOLLENBERG et al., 1970;
PEDERSEN and LADEFOGED, 1973).

B. Experimental Acute Renal Failure

The changes in renal hemodynamics occurring in acute renal failure
have been further delineated using experimental models. For example,
the renal hemodynamic alterations in uranyl nitrate-induced acute
renal failure have been demonstrated to parallel the alterations in
both total renal blood flow and the intrarenal distribution of blood
flow observed in clinical acute renal failure (FLAMENBAUM et al.,
1972a, 1974; KLEINMAN et al., 1975). Analyses of the renal hemodynamic and renovascular filling in glycerol-induced acute renal failure have also demonstrated a prompt fall in renal blood flow after the
initiation of acute renal failure (AYER et al., 1971; CHEDRU et al.,
1972). Thus, distinct parallels exist between renal hemodynamic alterations in clinical and experimental acute renal failure.

V. Synthesis

Diminished effective glomerular filtration rate, resulting in azotemia
with or without oliguria, is the central renal function abnormality in
clinical and experimental acute renal failure. The intensity of the
initiating event may telescope the development and course of acute
renal failure, and the net reduction in total kidney glomerular filtration rate is a reflection of the sum total of alterations in single
nephron filtration rates. The alterations in single nephron function
are, characteristically, heterogeneous, and run the gamut from absent
function to nearly normal function. The "common denominator" of the
initiating phase of acute renal failure appears to be an abnormality
in tubular function characterized by alterations in the transport of
sodium chloride and reabsorption of tubular fluid. Regardless of
whether this tubular dysfunction results from either a nephrotoxin or
from circulatory alterations with renal hypoperfusion, the alterations

in tubular fluid delivered to distal nephron segments results in an
increase in renin release from intrarenal sites, via stimulation of
the macula densa. The subsequent local generation of angiotensin, and
glomerular arteriolar vasomotion,cause decreases in glomerular plasma
flow, net ultrafiltration pressure and single nephron filtration. Any
contribution that an overflow of released renin may add to these
changes is minor. Thus, the mechanism of tubuloglomerular feedback
interrelates changes in tubular function (which depend primarily on
the integrity of actively metabolizing epithelial cells) to alterations in glomerular filtration rate (which depend primarily on the
physical force contributed by the cardiovascular system), and may be
considered the "final common mechanism" of acute renal failure. The
redistribution of intrarenal blood flow is the "final common pathway"
in acute renal failure and is a consequence of the physiologic response to tubular dysfunction.

That the abnormalities in active transport processes responsible for
the alterations in tubular fluid and electrolyte transport may be
reversible is suggested by the effect of dithiothreitol on uranyl
nitrate-induced alterations in turtle urinary bladder transport.
Based on these *in vitro* observations (see above), the effectiveness of
dithiothreitol reversal of heavy metal-induced alterations in membrane
function was evaluated *in vivo*. Alterations in glomerular filtration
as estimated from the endogenous clearance of creatinine, sodium excretion, and the renin-angiotensin system were evaluated in rats
receiving only uranyl nitrate (10 mg/kg body weight, subcutaneously)
alone, as well as in rats treated with dithiothreitol (15.4 mg/kg body
weight, intraperitoneally) 30 minutes after the administration of
uranyl nitrate. The results of these studies are presented in Table 3.
When rats received uranyl nitrate alone, there were marked increases
in plasma creatinine concentration, decreases in endogenous creatinine
clearance, and an increase in the fractional excretion of filtered

Table 3. Effect of dithiothreitol (DTT), 15.4 mg/kg i.p., on uranyl nitrate
(UN)-induced alterations in plasma creatinine (P_{Cr}), creatinine clearance
(C_{Cr}), fractional excretion of sodium ($FE_{Na}+$), superficial (S) and deep (D)
juxtaglomerular apparatus renin activity (JGA-RA) at 24 h after induction
of acute renal failure

Experimental Group	Control	DTT alone	UN alone	DTT + UN
P_{Cr} - mg%	0.46 ± 0.03[a]	0.44 ± 0.02	0.89 ± 0.06[b]	0.55 ± 0.02[b]
C_{Cr} - ml/min	1.27 ± 0.06	1.26 ± 0.06	0.50 ± 0.05[b]	0.93 ± 0.06[b]
$FE_{Na}+$ - %	0.77 ± 0.11	0.91 ± 0.09	2.63 ± 0.25[b]	1.24 ± 0.17
JGA-RA - ng/JGA/h				
S	24.90 ± 0.40	21.60 ± 1.60	37.20 ± 1.30[b]	17.50 ± 1.60
D	5.90 ± 0.20	7.50 ± 1.20	12.30 ± 0.60[b]	7.20 ± 1.00
PRA - ng/ml/h	5.30 ± 0.30	5.40 ± 1.20	10.00 ± 0.60[b]	4.50 ± 0.70

[a]Values presented are for mean ± SEM. There were 5-7 rats in each study
group.
[b]Significantly different from control value at 5% confidence limits or
below.

sodium, 24 hours after induction of acute renal failure. In a separate group of similarly handled rats, increased juxtaglomerular and plasma renin activities were observed during the same time interval. In contrast, rats receiving uranyl nitrate and dithiothreitol demonmonstrated significantly less severe alterations in these parameters. In particular, the alterations in sodium excretion and renin-angiotensin system activity were not observed, and the degree of decrease in glomerular filtration markedly diminisned. Thus, dithiothreitol not only reverses heavy metal-induced membrane dysfunction *in vitro*, but also significantly ameliorates the analogous pattern of events *in vivo*.

References

AYER, G., GRANDCHAMP, A., WYLER, J., TRUNIGER, B.: Intrarenal hemodynamics in glycerol-induced myohemoglobinuric acute renal failure in the rat. Circ. Res. 29, 128-135 (1971).

BAINES, A.D.: Redistribution of nephron function in response to chronic and acute saline loads. Am. J. Physiol. 224, 237-244 (1973).

BARENBERG, R.L., SOLOMON, S., PAPPER, S., ANDERSON, R.: Clearance and micropuncture study of renal function in mercuric chloride treated rats. J. Lab. Clin. Med. 72, 473-484 (1968).

BARRATT, L.J., WALLIN, J.D., RECTOR, F.C., SELDIN, D.W.: Influence of volume on single nephron filtration rate and plasma flow in the rat. Am. J. Physiol. 224, 643-650 (1973).

BARRON, E.S.G., MUNTZ, J.A., GASVODA, B.: Regulatory mechanisms of cellular respiration. I.: The role of cell membranes; uranium inhibition of cellular respiration. J. Gen. Physiol. 32, 163-178 (1948).

BEEUWKES, R.: Efferent vascular patterns and early vascular-tubular relations in the dog. Am. J. Physiol. 221, 1361-1374 (1971).

BEEUWKES, R., BONVENTRE, J.: Tubular organization and vascular-tubular relations in the dog. Am. J. Physiol. 229, 695-713 (1975).

BIBER, T.U., MYLLE, M., BAINES, A.D., GOTTSCHALK, C.W.: A study by micropuncture and microdissection of acute renal damage in rats. Am. J. Med. 44, 664-705 (1968).

BRENNER, B.M., FALCHUK, K.W., KEIMOWITZ, R.I., BERLINER, R.W.: The relationship between peritubular capillary protein concentration and fluid reabsorption by the renal proximal tubule. J. Clin. Invest. 48, 1519-1531 (1969).

BRENNER, B.M., TROY, J.L., DAUGHARTY, T.M., DEEN, W.M., ROBERTSON, C.R.: Dynamics of glomerular ultrafiltration in the rat. II.: Plasma-flow dependence of GFR. Am. J. Physiol. 223, 1184-1190 (1972).

BROWN, J.J., GLEADLE, R.I., LAWSON, D.H., LEVER, A.F., LINTON, A.L., MacADAM, R.F., PRENTICE, E., ROBERTSON, J.I.S., TREE, M.: Renin and acute renal failure: Studies in man. Brit. J. Med. 1, 253-258 (1970).

BROWN, J.J., CHINN, R.H., GAVRAS, H., LECKE, B., LEVER, A.F., MacGREGOR, J., MORTON, J., ROBERTSON, J.I.S.: Renin and renal function. In: Hypertension (Eds. J. GENEST, E. KARUV). Berlin: Springer 1972.

BROWN, W.C.B., BROWN, J.J., GAVRAS, H., POWELL-JACKSON, J.D., LEVER, A.F., MacGREGOR, J., MacADAM, R.F., ROBERTSON, J.I.S.: Renin and acute circulatory renal failure in the rabbit. Circ. Res. <u>30</u>, 114-122 (1972).

BRUNS, F.J., ALEXANDER, E.A., RILEY, A.L., LEVINSKY, N.G.: Superficial and juxtamedullary nephron function during saline loading in the dog. J. Clin. Invest. <u>53</u>, 971-979 (1974).

BULL, G.M., JOEKES, A.M., LOWE, K.G.: Renal function studies in acute tubular necrosis. Clin. Sci. <u>9</u>, 379-403 (1950).

BUNAG, R.D., PAGE, I.H., McCUBBIN, W.J.: Neural stimulation of renin release. Circ. Res. <u>19</u>, 851-858 (1966).

CHEDRU, M.F., BAETHKE, R., OKEN, D.E.: Renal cortical blood flow and glomerular filtration in myohemoglobinuric acute renal failure. Kidney Int. <u>1</u>, 232-239 (1972).

CLELAND, W.W.: Dithiothreitol, a new protective agent for SH groups. Biochemistry <u>3</u>, 480-482 (1964).

CLEMENT, D.L., PELLETIER, C.L., SHEPERD, J.T.: Role of vagal afferents in the control of renal sympathetic activity in the rabbit. Circ. Res. <u>31</u>, 824-830 (1972).

COOK, W.F.: Cellular localization of renin. In: Kidney Hormones (Ed. J.W. FISCHER), New York: Little, Brown 1971.

CORTNEY, M.A., NAGLE, W., THURAU, K.: A micropuncture study of the relationship between flow rate through the loop of Henle and sodium concentration in the early distal tubule. Pflug. Arch. <u>287</u>, 286-295 (1966).

DAHLHEIM, H., GRANGER, P., THURAU, K.: A sensitive method for determination of renin activity in the single juxtaglomerular apparatus of the rat kidney. Pflug. Arch. Europ. J. Physiol. <u>321</u>, 303-315 (1970).

DAUGHARTY, T.M., VEKI, I.F., MERCER, P.F., BRENNER, B.M.: Dynamics of glomerular filtration in the rat. V.: Response to ischemic injury. J. Clin. Invest. <u>53</u>, 105-116 (1974).

DAVIS, J.O.: What signals the kidney to release renin? Circ. Res. <u>28</u>, 301-306 (1971).

DEEN, W.M., ROBERTSON, C.R., BRENNER, B.M.: Transcapillary fluid exchange in the renal cortex. Circ. Res. <u>33</u>, 1-8 (1973).

DEMIS, D.J., ROTHENSTEIN, A., MEIER, R.: The relationship of the cell surface to metabolism. X.: The location and function of invertase in the yeast cell. Arch. Biochem. <u>48</u>, 55-62 (1954).

DEV, B., DRESCHER, C., SCHNERMANN, J.: Resetting of tubuloglomerular feedback sensitivity by the dietary salt intake. Pflug. Arch. <u>346</u>, 263-277 (1974).

DiBONA, G.F., McDONALD, F., FLAMENBAUM, W., DAMMON, G.J., OKEN, D.E.: Maintenance of renal function in salt loaded rats despite severe tubular necrosis induced by $HgCl_2$. Nephron <u>8</u>, 205-220 (1971).

DiBONA, G.F., SAWIN, L.L.: The renin-angiotensin system in acute renal failure in the rat. Lab. Invest. <u>25</u>, 528-532 (1971).

FINKIELMAN, S., GOLDSTEIN, D.J., FISCHER-FENARO, C., NAHNOL, V.F.: *In vitro* production of angiotensin and renin release by isolated glomeruli. Medicine (Buenos Aires) <u>32</u>, (suppl I) 37-39 (1972).

FINN, A.L.: Separate effects of sodium and vasopressin on the sodium pump in toad bladder. Am. J. Physiol. <u>215</u>, 849-856 (1968).

FLAMENBAUM, W.: Pathophysiology of acute renal failure. Arch. Intern. Med. (Chicago) <u>131</u>, 911-928 (1973).

FLAMENBAUM, W., HAMBURGER, R.J.: Superficial and deep juxtaglomerular apparatus renin activity of the rat kidney: Effect of surgical preparation and NaCl intake. J. Clin. Invest. <u>54</u>, 1373-1381 (1974).

FLAMENBAUM, W., HAMBURGER, R.J., HUDDLESTON, M.L., KAUFMAN, J., McNEIL, J.S., SCHWARTZ, J.H., NAGLE, R.: The initiation phase of experimental acute renal failure: an evaluation of uranyl nitrate-induced acute renal failure in the rat. Kidney Int. (in press).

FLAMENBAUM, W., HUDDLESTON, M.L., McNEIL, J.S., HAMBURGER, R.J.: Uranyl nitrate-induced acute renal failure in the rat: micropuncture and renal hemodynamic studies. Kidney Int. 6, 408-418 (1974).

FLAMENBAUM, W., KOTCHEN, T.A., OKEN, D.E.: Effect of renin immunization on mercuric chloride and glycerol-induced renal failure. Kidney Int. 1, 406-412 (1972a).

FLAMENBAUM, W., McDONALD, F.D., DiBONA, G.F., OKEN, D.E.: Micropuncture study of renal tubular factors in low dose mercury poisoning. Nephron 8, 221-234 (1971).

FLAMENBAUM, W., McNEIL, J.S., KOTCHEN, T.A., LOWENTHAL, D., NAGLE, R.B.: Glycerol-induced acute renal failure after plasma renin activity suppression. J. Lab. Clin. Med. 82, 587-596 (1973).

FLAMENBAUM, W., McNEIL, J.S., KOTCHEN, T.A., SALADINO, A.J.: Experimental acute renal failure induced by uranyl nitrate in the dog. Circ. Res. 31, 682-698 (1972b).

FLANNIGAN, W.J., OKEN, D.E.: Renal micropuncture study of the development of anuria in the rat with mercury induced acute renal failure. J. Clin. Invest. 44, 449-457 (1965).

FRENKEL, A., EKBALD, E.B.M., EDELMAN, I.S.: Effects of sulfhydryl reagents on basal and vasopressin-stimulated Na^+ transport in the toad bladder. In: Biomembranes 7, (Eds. H. EISENBERG, E. KATACHELSKI-KATZIR, L.A. MANSON). New York: Plenum Press 1975.

GAVRAS, H., BROWN, J.J., LEVER, A.F., ROBERTSON, J.I.S.: Changes of renin in individual glomeruli in response to variations of sodium intake. Clin. Sci. 38, 409-414 (1970).

GILMORE, J.P.: Contribution of baroreceptors to the control of renal function. Circ. Res. 14, 301-317 (1964).

GOORMAGHTIGH, N.: Histologic changes in the ischemic kidney, with special reference to the juxtaglomerular apparatus. Am. J. Pathol. 16, 409-416 (1940).

GOORMAGHTIGH, N.: Vascular and circulatory changes in renal cortex in anuric crush-syndrome. Proc. Soc. Exp. Biol. Med. 59, 303-305 (1945).

GOORMAGHTIGH, N.: The renal arteriolar changes in the anuric crush syndrome. Am. J. Pathol. 23, 513-529 (1947).

GRANGER, P., DAHLHEIM, H., THURAU, K.: Enzyme activity of single juxtaglomerular apparatuses in the rat kidney. Kidney Int. 1, 77-88 (1972).

HENRY, L.N., LANE, C.E., KASHGARIAN, M.: Micropuncture studies of the pathophysiology of acute renal failure in the rat. Lab. Invest. 19, 309-314 (1968).

HIRSCHORN, N., FRAZIER, H.S.: Intracellular electrical potential of the epithelium of turtle bladder. Am. J. Physiol. 220, 1158-1161 (1971).

HOLLENBERG, N., ADAMS, D.F., OKEN, D.E., ABRAMS, H.L., MERRILL, J.P.: Acute renal failure due to nephrotoxins. Renal hemodynamics and angiographic studies in man. New Engl. J. Med. 282, 1329-1334 (1970).

HOLLENBERG, N.K., EPSTEIN, M., ROSEN, M., BASON, R.R., OKEN, D.E., MERRILL, J.P.: Acute oliguric renal failure in man: evidence for preferential renal cortical ischemia. Medicine (Baltimore) 47, 455-474 (1968).

112

JAENIKE, J.R.: Micropuncture study of methemoglobin-induced renal failure in the rat. J. Lab. Clin. Med. 73, 459-468 (1969).

JAMISON, R.L.: Micropuncture study of superficial and juxtamedullary nephrons in the rat. Am. J. Physiol. 218, 46-55 (1970).

JAMISON, R.L., LACY, F.B.: Effect of saline infusion on superficial and juxtamedullary nephrons in the rat. Am. J. Physiol. 221, 690-697 (1971).

JOHNSON, J.A., DAVIS, J.O., WITTY, R.T.: Effects of catecholamines and renal nerve stimulation on renin release in the non-filtering kidney. Circ. Res. 29, 646-653 (1971).

KEZDI, P., GELLER, E.: Baroreceptor control of post-ganglionic sympathetic nerve discharge. Am. J. Physiol. 214, 427-435 (1968).

KLEINMAN, J.G., McNEIL, J.S., FLAMENBAUM, W.: Uranyl nitrate acute renal failure in the dog: early changes in renal function and hemodynamics. Clin. Sci. Mol. Med. 48, 9-19 (1975).

LADEFOGED, J., WINKLER, K.: Hemodynamics and acute renal failure. The effect of hypotension induced by hydralazine on renal blood flow, mean circulation time for plasma, and renal vascular volume in patients with acute oliguric renal failure. Scand. J. Clin. Lab. Invest. 26, 83-87 (1970).

LAGRANGE, R.G., SLOOP, C.H., SCHMIDT, H.B.: Selective stimulation of renal nerves in the anesthetized dog. Effect on renin release during controlled changes in renal hemodynamics. Circ. Res. 33, 704-712 (1973).

LARAGH, J.H., SEALEY, J.E.: The renin-angiotensin-aldosterone hormonal system and regulation of sodium, potassium and blood pressure homeostasis. In: Handbook of Physiology (Eds. J. ORLOFF, R.W. BERLINER), Sect. VIII. Baltimore: Williams and Wilkins Co. 1973.

LESLIE, B.R., SCHWARTZ, J.H., STEINMETZ, P.R.: Coupling between Cl- absorption and HCO_3^- secretion in turtle urinary bladder. Am. J. Physiol. 225, 610-617 (1973).

LEVER, A.F., PEART, W.S.: Renin and angiotensin-like activity in renal lymph. J. Physiol. (Lond.) 160, 548-563 (1972).

LICHTENSTEIN, N.F., LEAF, A.: Effect of amphotericin B on the permeability of toad bladder. J. Clin. Invest. 44, 1328-1342 (1969).

LJUNGQVIST, A.: Ultrastructural demonstration of a connection between afferent and efferent juxtaglomerular arterioles. Kidney Int. 8, 239-244 (1975).

LJUNGQVIST, A., WAGERMARK, J.: Adrenergic innervation of intrarenal glomerular and extraglomerular circulatory routes. Nephron 7, 218-229 (1970).

LOGAN, A., JOSE, P., EISNER, G., LILLIENFIELD, L., SLOTKOFF, L.: Intracortical distribution of renal blood flow in hemorrhagic shock in dogs. Circ. Res. 29, 257-266 (1971).

MASSANI, Z.M., FINKIELMAN, S., WORCEL, M., AGREST, A., PALADINI, A.C.: Angiotensin blood levels in hypertensive and nonhypertensive disease. Clin. Sci. 30, 473-483 (1966).

MATHEWS, P.G., MORGAN, T.O., JOHNSTON, C.I.: The renin-angiotensin system in acute renal failure in rats. Clin. Sci. Mol. Med. 47, 79-88 (1974).

McDONALD, F.D., THIEL, G., WILSON, R.D., DiBONA, G.F., OKEN, D.E.: The prevention of acute renal failure in the rat by long term saline loading. A possible role of the renin-angiotensin axis. Proc. Soc. Exp. Biol. Med. 131, 610-614 (1969).

MENDELSOHN, F.A.O., JOHNSTON, C.I.: Composition of juxtaglomerular
   granules. *In vitro* presence of angiotensin immunoactive substance.
   Proc. 5th Int. Cong. Nephrol. Mexico City, 1972.
MERRILL, J.P.: Kidney disease: acute renal failure. Ann. Rev. Med.
   11, 127-150 (1960).
MERRILL, J.P.: Acute renal failure. In: Disease of the Kidney (Eds.
   M.B. STRAUSS, L.G. WELT), 2nd ed., pp. 637-666. Boston: Little,
   Brown and Co. 1971.
MOGIL, R.A., ITSKOWITZ, H., RUSSELL, J.H., MURPHY, J.J.: Renal
   innervation and renin activity in salt metabolism and hypertension.
   Am. J. Physiol. 216, 693-697 (1969).
NAVAR, L.G., BURKE, T.J., ROBINSON, R.R., CLAPP, J.R.: Distal tubular
   feedback in the autoregulation of single nephron filtration rate.
   J. Clin. Invest. 53, 516-525 (1974).
NINOMIYA, I., IRISAWA, H.: Summation of baroreceptor reflex effects
   on sympathetic nerve activity. Am. J. Physiol. 216, 1330-1336 (1969).
NINOMIYA, I., NISIMARU, N., IRISAWA, H.: Sympathetic nerve activity
   to the spleen, kidney and heart in response to baroreceptor input.
   Am. J. Physiol. 221, 1346-1351 (1971).
NORBY, L.E., AL-AWQATI, Q., MULLER, A., STEINMETZ, P.R.: Separate
   stimulation of H^+ secretion and Na^+ absorption by aldosterone in
   turtle bladder. Clin. Res. 23, 371A (1975).
OCHOA, E., FINKIELMAN, S., AGREST, A.: Angiotensin blood levels during
   the evolution of acute renal failure. Clin. Sci. 38, 225-231 (1970).
OKEN, D.E., ARCE, M.L., WILSON, D.R.: Glycerol-induced hemoglobinuric
   acute renal failure in the rat. I.: Micropuncture study of the
   development of oliguria. J. Clin. Invest. 45, 724-735 (1966).
OKEN, D.E., COATES, S.C., FLAMENBAUM, W., POWELL-JACKSON, J.D., LEVER,
   A.F.: Active and passive immunization to angiotensin in experimental
   acute renal failure. Kidney Int. 7, 12-18 (1975a).
OKEN, D.E., MENDE, C.W., TARABA, I., FLAMENBAUM, W.: Resistance to
   acute renal failure by prior renal failure: examination of the role
   of renal renin content. Nephron 15, 131-142 (1975b).
OSBORNE, M.J., DROZ, B., MEYER, P., MOREL, F.: Angiotensin II: renal
   localization in glomerular mesangial cells by autoradiography.
   Kidney Int. 8, 245-254 (1975).
PAGE, I.H., McCUBBIN, J.W.: Renal Hypertension. Chicago: Yearbook
   Medical Publ. Inc. 1968.
PAPPER, S.: Renal failure. Med. Clin. N. Am. 55, 335-357 (1971).
PEDERSON, F., LADEFOGED, J.: Renal hemodynamics in acute renal failure
   in man measured by intra-arterial injection. External counting with
   Xenon-133 and I-131-albumin. Scand. J. Urol. Nephrol. 7, 187-195
   (1973).
POMERANZ, B.H., BIRTCH, A.G., BARGER, A.C.: Neural control of intra-
   renal blood flow. Am. J. Physiol. 215, 1067-1081 (1968).
POWELL-JACKSON, J.D., BROWN, J.J., LEVER, A.F., MacGREGOR, J.,
   MacADAM, R.F., TITTERINGTON, D.M., ROBERTSON, J.I.S., WAITE, M.A.:
   Protection against acute renal failure in rats by passive immunization against angiotensin II. Lancet 1, 774-776 (1972).
RAUGH, W.H., OSTER, P., DIETZ, R., GROSS, I.: Angiotensin II in acute
   renal failure in the rat. Acta Endocrinol. 73, (suppl 177) 193
   (1973).
REUBI, F.C., GOSSWEILER, N., GURTLER, R.: A dye dilution method of
   measuring renal blood flow in man with special reference to the
   anuric subject. Proc. Soc. Exp. Biol. Med. 111, 760-765 (1962).

REUSS, L., FINN, A.L.: Dependence of serosal membrane potential on mucosal membrane potential in toad urinary bladder. Biophys. J. 15, 71-75 (1975).

RIGGS, A.: The oxygen equilibrium of the hemoglobin of the eel. J. Gen. Physiol. 35, 41-44 (1952).

RIGGS, A.: Hemoglobin structure. In: Sulfur Protein (Ed. R. BENESCH), p. 173. New York: Academic Press 1959.

RIGGS, A., WALBACH, R.A.: Sulfhydryl groups and the structure of hemoglobin. J. Gen. Physiol. 39, 585-605 (1956).

ROTHSEIN, A.: Cell membrane as a site of action of heavy metals. Fed. Proc. 18, 1026-1035 (1959).

RUIZ-GUINAZU, A.: Alterations of the glomerular filtration rate in acute renal failure. In: Pathogenesis and Clinical Findings with Renal Failure (Eds. U. GESSLER, K. SCHROEDER, H. WEIDINGER), pp. 23-32. Stuttgart: Georg Thieme 1971.

RYAN, R., McNEIL, J.S., FLAMENBAUM, W.: Uranyl nitrate-induced acute renal failure in the rat: effect of varying doses and saline loading. Proc. Soc. Exp. Biol. Med. 143, 289-296 (1973).

SCHNERMANN, J., DAVIS, J.M., WUNDERLICH, P., LEVINE, D.Z., HORSTER, M.: Technical problems in the micropuncture determination of nephron filtration rate and their functional implications. Pflug. Arch. 329, 307-320 (1971).

SCHNERMANN, J. PERSSON, A.E.G., AGERUP, B.: Tubuloglomerular feedback. Nonlinear relation between glomerular hydrostatic pressure and loop of Henle perfusion rate. J. Clin. Invest. 52, 862-869 (1973).

SCHNERMANN, J., NAGEL, W., THURAU, K.: Die fruhdistale Natrium-konzentration in Rattenieren nach Renaler Ischamie und hemorrhagischer Hypotension. Ein Beitrag zur Pathogenese der postischamischen und posthemorrhagischen Filtraterniedrigung. Pflug. Arch. 287, 296-310 (1966).

SCHNERMANN, J., WRIGHT, F.S., DAVIS, J.M., STACKELBERG, W.V., GRILL, G.: Regulation of superficial nephron filtration rate by tubuloglomerular Pflug. Arch. 318, 147-175 (1970).

SCHWARTZ, J.H.: H^+ current response to CO_2 and carbonic anhydrase inhibitors by the turtle urinary bladder. Am. J. Physiol. (in press).

SCHWARTZ, J.H., FLAMENBAUM, W.: Heavy metal induced alterations in ion transport by the turtle urinary bladder. Am. J. Physiol. (in press).

SEVITT, S.: Pathogenesis of traumatic uremia. A revised concept. Lancet 2, 135-1141 (1959).

STEINHAUSEN, M., EISENBACH, G.M., HELMSTADLER, V.: Concentration of lissamine green in proximal tubules of antidiuretic and mercury poisoned rats. Pflug. Arch. 311, 1-15 (1969).

STEINMETZ, P.R.: Cellular mechanisms of urinary acidification. Physiol. Rev. 54, 890-956 (1974).

STEINMETZ, P.R., LAWSON, L.R.: Defect in urinary acidification induced by amphotericin B. J. Clin. Invest. 49, 596-601 (1970).

STEINMETZ, P.R., LAWSON, L.R.: Affect of luminal pH on ion perme-ability and flow of Na^+ and H^+ in turtle bladder. Am. J. Physiol. 220, 1573-1580 (1971).

STUMPE, K.O., LOWITZ, H.D., OCHWALD, B.: Function of the juxta-medullary nephrons in normotensive and chronically hypertensive rats. Pflug. Arch. Ges. Physiol. 313, 43-52 (1969).

SUTHERLAND, L.E.: Fluorescent antibody study of juxtaglomerular space cells using the freeze-substitution technique. Nephron 7, 512-523 (1970).

SWANN, R.C., MERRILL, J.P.: The clinical course of acute renal failure. Medicine (Baltimore) 32, 215-292 (1953).

TAKABATAKE, T., MASON, J., FLAMENBAUM, W., THURAU, K.: Micropuncture Investigation in Acute Renal Failure. Tenth Gesellschaft für Nephrologie, Innsbruck, 1974.

THIEL, G., McDONALD, F.D., OKEN, D.E.: Micropuncture studies of the basis for protection of renin depleted rats from glycerol-induced acute renal failure. Nephron 7, 67-79 (1970).

THORBURN, G.D., KOPALD, H.H., HERD, J.A., HOLLENBERG, N., O'MORCHOE, C.C.C., BARGER, A.C.: Intrarenal distribution of nutrient blood flow determined with Kr^{85} in the unanesthetized dog. Circ. Res. 13, 290-305 (1963).

THURAU, K.: Influence of sodium concentration at macula densa cells on tubular sodium load. Ann. N.Y. Acad. Sci. 139, 388-399 (1966).

THURAU, K.: Pathophysiologie des akuten Nierenversagen. In: Anaesthesiologie und Wiederbelegung. Vol. 49, Intensivtherapie beim akuten Nierenversagen, pp. 1-18. Berlin: Springer-Verlag 1970.

THURAU, K.: The juxtaglomerular apparatus: its role in the function of the single nephron unit. In: Modern Diuretic Therapy in the Treatment of Cardiovascular and Renal Disease. Int. Cong. Series No. 268, pp. 84-93. Amsterdam: Excerpta Medica 1972a.

THURAU, K.: Aspects in renal physiology. Klin. Wschr. 50, 221-225 (1972b).

THURAU, K.W.C., DAHLHEIM, H., GRUNER, A., MASON, J., GRANGER, P.: Activation of renin in the single juxtaglomerular apparatus by sodium chloride in the tubular fluid at the macula densa. Circ. Res. 30-31, (suppl II) 182-186 (1972).

THURAU, K., MASON, J.: The intrarenal function of the juxtaglomerular apparatus. In: Kidney and Urinary Tract Physiology, MTP International Review of Science, Physiology Ser. 1, vol. VI, pp. 357-390. London: Butterworth and Co. 1974.

THURAU, K., SCHNERMANN, J.: Die Natriumkonzentration an den Macula densa-Zellen als regulierender Faktor fur das Glomerulum filtrat (Mikropunktionversuche). Klin. Wschr. 43, 410-413 (1965).

TU, W.H.: Plasma renin activity in acute tubular necrosis and other renal diseases associated with hypertension. Circulation 31, 686-695 (1965).

WALLIN, J.D., BLANTZ, R.C., KATZ, M.A., ANDREUCCI, V.E., RECTOR, F.C., SELDIN, D.W.: Effect of saline diuresis on intrarenal blood flow in the rat. Am. J. Physiol. 221, 1297-1304 (1971).

WEBB, J.L.: Enzyme and Metabolic Inhibitors, vol. II, p. 1237. New York: Academic Press 1966.

WIRZ, H.R., DIRIX, R.: Urinary concentration and dilution. In: Handbook of Physiology, Sect. VIII (Eds. J. ORLOFF, R.W. BERLINER). Baltimore: Williams and Wilkins Co. 1973.

WRIGHT, F.S., SCHNERMANN, J.: Interference with feedback control of glomerular filtration rate by furosemide, triflocin and cyanide. J. Clin. Invest. 53, 1695-1708 (1974).

ZEHR, J.E., FEIGL, E.O.: Suppression of renin activity by hypothalamic stimulation. Circ. Res. 32, (suppl I) 17-27 (1973).

Three-Dimensional Pharmacophoric Pattern Searching

Peter Gund

I. Introduction

The medicinal chemist has long known that modifying a drug's chemical structure may cause modified biologic response. If we accept the view that biologic reactions are subject to normal chemical forces, then the usual physical chemical concepts—e.g., thermodynamic and kinetic theory—should apply. However, early attempts to correlate biologic activity quantitatively with chemical structure, using the Hammett equation or Hückel molecular orbital calculations, met with only limited success (GOODFORD, 1973; PURCELL et al., 1973). It is now clear that this failure was due to the greater complexity of bioreactions compared to "normal" chemical reactions in homogeneous solution.

A. Differences between Bioreactions and Solution Reactions

It has proved difficult to find a consistent and valid measure of biologic action, due to the variable response of individual test subjects, and the problem of devising tests in which the measured response is simply and unequivocally related to drug dose (GOURLEY, 1970; GOODFORD, 1973). Drug may be transported to site of action by passive diffusion and partitioning through various membranes and fluids, by active (energy-coupled) transport, or by facilitated transport or selective diffusion through a channel (GOURLEY, 1970; WILBRANDT and ROSENBERG, 1961). Drug may bind to plasma or "false receptors"; it may be rapidly excreted or inactivated; or it may encounter such mysterious obstacles as the blood-brain barrier (OLDENDORF, 1974). Metabolic transformation (McMAHON, 1970) may increase as well as decrease drug potency (SINKULA and YALKOVSKY, 1975).

Biologic processes generally are enzyme-mediated and are subject to greater stereochemical constraints on reagent structure than the corresponding solution reactions (CASY, 1970; HANSON and ROSE, 1975). Although solution reactions usually may proceed to completion (thermodynamic equilibrium), total bioequilibrium is reached only after the organism has died. In order to maintain the chemical driving force necessary for life itself, bioreactions exhibit extremely complex kinetics, involving feedback control loops and profoundly interdependent reaction pathways (CITRI, 1973; SEGAL, 1973). Nature appears to go to great lengths to make biochemical processes discontinuous; a cell process (e.g., mitosis, muscle contraction, nerve stimulus transmission) tends to be in an "off" state until "turned on" by what may be viewed as a "pattern recognition" response to an activating molecule or stimulus.

Despite all these complications—and more—in biologic processes, quantitative structure-activity relationships (QSARs) are often found for a family of drugs.

II. Quantitative Structure-Activity Relationships

QSARs are derived by means of a number of statistical techniques, which may broadly be categorized as multiple regression (MR) or pattern recognition (PR) analysis.

A. Multiple Regression (MR) Analysis

This well-founded empirical procedure applies standard regression techniques to the problem of correlating activity with several possible structural parameters (REDL et al., 1974; HANSCH, 1973; PURCELL et al., 1973; GOODFORD, 1973). Three major subcategories are *de novo* correlations, linear free energy relationships, and substructural analysis.

The FREE-WILSON (1964) *de novo* analysis assumes that substituents on a drug parent system influence activity independently of each other. In this procedure, a substituent (say, *para*-chloro) is assigned a contribution of 1 to the activity when it is present, 0 when absent. The resulting simultaneous equations (equated to observed activity) are solved by least squares techniques, and the derived coefficient indicates whether a given substituent enhances or hinders activity. Where correlations are found, the method offers the possibility of predicting activity of compounds with all combinations and permutations of the given substituents.

The linear free energy relationship (LFER) approach (WELLS, 1968; SHORTER, 1973) assigns known physicochemical parameters to the substituents and again solves simultaneous equations. This approach has been utilized extensively to relate the rate and equilibrium of solution reactions to chemical structure since it was introduced by HAMMETT (1937). TAFT and LEWIS (1959) reformulated LFER's as a multiparameter equation, and HANSCH et al. (1962) was able to extend the method successfully to biologic reactions. The empirical technique is based on the "extrathermodynamic" assumption—i.e., the assumption is not justified by thermodynamic theory (LEFFLER and GRUNWALD, 1963)—that the free energy of a reaction can be partitioned into a linear sum of individual energetic contributions. Although many variations of the method have been published, and different research groups have "favorite" substituent constants that they try to correlate, the normal analysis includes an electronic term, a partitioning term, and perhaps a steric term (HANSCH et al., 1973a; NORRINGTON et al., 1975).

The electronic term is traditionally HAMMETT's sigma substituent constant (JOHNSON, 1973; SHORTER, 1973), although this is often subdivided into field and resonance terms [$\mathcal{F}$ and $\mathcal{R}$ in the SWAIN-LUPTON (1968) approach]. Partition coefficient is usually represented by the HANSCH pi constant (FUJITA et al., 1964), often with a squared term added to reflect the observed hyperbolic relationship of activity and hydrophobicity (extremely lipid soluble or extremely water soluble compounds have difficulty traversing all biophases to reach site of action: HANSCH, 1969).

Expressing the steric effect in an LFER has long been a problem (CAVALLITO, 1973; SHORTER, 1972; LIEN, 1969; NORRINGTON et al., 1975; SIMON, 1974). Solution reactions usually have been correlated with TAFT's steric constant E_s, but van der Waals radius (CHARTON, 1975), molar refractivity and molecular weight (HANSCH et al., 1973a; NORRINGTON et al., 1975), orientation constants (ÖTVÖS et al., 1975), topologic indices (DUBOIS et al., 1973), calculated congestion (WIPKE and GUND, unpublished results), principal molecular axes or moments of gyration (ZANDER and JURS, 1975), three-dimensional Fourier transform (DIERDORF and KOWALSKI, 1974), and model hydrocarbon strain calculations (DETAR, 1974a,b)—to name just a few recent examples—have also been used as measures of steric environment. Nevertheless, correlation of bioactivity with steric effects has been accomplished with only a few of these indices, and in a limited number of cases (e.g., HANSCH, 1973). This is perhaps not surprising, given the strong influence of reagent geometry on enzymic reactions (HANSON and ROSE, 1975).

Recently, substructural analysis has been added to the arsenal of MR techniques. CRAMER et al. (1974) were partially successful in correlating drug activity with molecular substructure as represented by a fragment code. In a similar approach, ADAMSON and BUSH (1974) correlated antibacterial activity of 79 penicillins with substructural fragments. This technique has also been used recently in an attempt to elucidate chemical structure from ^{13}C nmr spectra (BREMSER et al., 1975). In an attempt to correlate activity with three-dimensional steric environment, SIMON and SZABADAI (1973) defined a minimum steric difference (MSD) parameter essentially as the number of main row atoms remaining when a model was constructed to overlap as many atoms of the reference compound as possible. The correlation of MSD with activity, again, was encouraging but not overwhelming.

B. Pattern Recognition Analysis

The rapidly expanding field of artificial intelligence includes pattern recognition (PR) techniques (JURS and ISENHOUR, 1975; KOWALSKI, 1974; KOWALSKI and BENDER, 1975) such as learning machines, k-nearest neighbor calculations, discriminant analysis, and many others. PR techniques have been used to correlate biologic activity with mass spectral fragmentation patterns (TING et al., 1973) and with molecular fragments. Among others, HILLER et al. (1973), STUPER and JURS (1975), CHU (1974), CHU et al. (1975), and KOWALSKI and BENDER (1974) have "recognized" molecular patterns that appear to be associated with

bioactivity. While PR techniques may be used with any numeric information—including physicochemical parameters—it has largely been
used to date to correlate bioactivity with molecular fragement information.

C. QSAR Generalizations

MR techniques hinge on the tenuous assumption that variation of chemical structure with bioactivity is describable by a smooth, continuous
curve. While this is doubtless not totally true, the method works
surprisingly well in many and varied cases for predicting optimally
active examples of a homogeneous set of compounds. Nevertheless, an
analog containing a substituent with different steric properties,
which is metabolized, or which changes site of binding can fail to
correlate.

PR techniques are capable of handling activity discontinuities and
inactive analogs. In principle, they can correlate activity among
highly heterogeneous sets of compounds. In practice, however, the
method appears to contain pitfalls, and published analyses have been
the subject of controversy. For example, the correlation of bioactivity with mass spectral data (TING et al., 1973) has been called
trivial (PERRIN, 1974) and absurd (CLERC et al., 1973); and a correlation of anticancer data with structural fragments (KOWALSKI and
BENDER, 1974) has been criticized as valid only for the highly biased
set of compounds used in the analysis (MATHEWS, 1975). It has been
said that there is a danger with PR techniques of correlating insignificant information (KENT and GAÜMANN, 1975). PR results seem to be
highly dependent upon patterns chosen for input; and since a very
large number of patterns are generally possible, a fair amount of
subjective judgment enters into the analysis. Further problems include pattern overlap and redundancy and relative weighting of patterns.
Nevertheless, PR has the capability of predicting new classes of drugs
with a desired activity, and deserves to be pursued. It is likely
that the most satisfactory results will be attained when fairly large
substructures are used to provide good discrimination.

Both MR and PR techniques are parametric methods that are subject to
dataset bias and to choice of independent parameters. Furthermore,
the specification of stereochemical and spatial information in a
general way has proved difficult with either technique. The remainder
of this chapter develops the concept of a pharmacophoric pattern as a
representation of structural information that might be correlated with
bioactivity by either MR or PR methods.

III. Description of Pharmacophoric Patterns

We define a *pharmacophoric pattern* as *a collection of atoms spatially
disposed in a manner that elicits a biologic response*. The importance
of spatial configuration is a consequence of the drug-receptor theory

of biologic action, which is discussed below. Since both topologic and topographic patterns have been discussed in the literature, we will consider them separately. Historically, the term "pharmacophore" was attributed to EHRLICH (1909) by ARIËNS (1966), and the distinction between topologic and topographic patterns has been recognized by KAUFMAN and KERMAN (1974), among others.

A. Topologic Molecular Patterns

Topologic patterns are independent of molecular conformation, but dependent upon molecular configuration (positional isomerism). They are specified in terms of atom type and connectivity, as contained in a connection table representation of structure (LYNCH et al., 1971). They correspond to molecular fragments or substructures. They have been defined "graph theoretically" as atom-centered or bond-centered fragments (LYNCH, 1974), or as "environments" (DUBOIS, 1974). They have been defined chemically in terms of functional groups, rings, side chains, and Wiswesser Line Notation symbols (EAKEN and HYDE, 1974; FELDMAN, 1974). These types of patterns have been used extensively in pattern recognition studies of influence of molecular structure on bioactivity.

B. Topographic Molecular Patterns

Topographic patterns are geometry dependent, conformation dependent, and specified by atom type and position in space. Their importance for bioactivity is a consequence of the drug-receptor theory of biologic action (CAVALLITO, 1973; KOROLKOVAS, 1970).

In essence, receptor theory states that a drug must interact chemically with an appropriate site on a biologic macromolecule in order to exert its effect. ("Drug" is used throughout in its larger sense, as any chemical—synthetic or natural—that causes a biologic effect.) As the drug complexes with the receptor site, a biologic response (perhaps due to a macromolecular conformational change) occurs, and the drug is released, allowing the biomolecule to become primed for the next activating episode. Drug action may be categorized as agonist—eliciting a normal biologic response from the receptor—and antagonist—inhibiting normal response by the receptor. Antagonist action may occur by strong or irreversible complexing to the receptor site, or by binding to some other site on the biomolecule, so normal binding to the distorted receptor site is not possible (allosteric inhibition: GRAY, 1971).

C. Pharmacophoric Patterns: Topologic or Topographic?

Bioreceptors are believed to possess a specific geometric structure. Therefore, the receptor site pattern, and the complementary pharmacophoric pattern, are topographic in nature. In recognition of this principle, CROXATTO and HUIDOBRO (1956) molded receptor sites by "surface complementarity" to space-filling models of active drugs in order to predict bioactivity.

Nevertheless, a number of topologic pharmacophoric patterns have been proposed in the literature and have often proved useful in classifying drugs of similar activities. *Topologic patterns will succeed in describing bioactivity insofar as they correspond to the actual (topographic) pharmacophoric pattern.* Thus, patterns involving unsaturated rings or constrained rings possess relatively few degrees of conformational freedom, and the geometric structure is often inherent in the structural diagram. Chiral topologic patterns contain additional spatial information. Furthermore, distance between functional groups (topography) may be approximately represented as a number of single bonds (topology). For example, JANSSEN (1973) proposed that the basic topologic pattern shown in 1, where X may be several different atom types,

$$Ph-X-X-X-X-NR_2$$

<u>1</u>

may be associated with antipsychotic activity. This is clearly a simplification, since many compounds containing that substructure are not neuroleptics. JANSSEN (1970) proposed an S-shaped topography for this pattern, and quantum mechanical calculations (KAUFMAN and KERMAN, 1974; KAUFMAN and KOSKI, 1975) and classical calculations (FEINBERG and SNYDER, 1975) have shown that these drugs do tend to assume certain well-defined conformations. Furthermore, in the preferred conformation, at least some of them may be superimposed on the preferred conformation of dopamine, rationalizing their competitive binding to the same receptor (FEINBERG and SNYDER, 1975). The dynamics of conformational change in these drugs has also been related to antipsychotic activity (FENNER, 1974).

The presence of a topographic pharmacophoric pattern is, of course, no guarantee of activity, since transport, metabolism, and steric problems may interfere. The further complication of molecular flexibility will be considered below. Nonetheless, it appears that the most discriminating measure of potential bioactivity will be an appropriate topographic pattern.

D. Functional Definition of Pharmacophoric Pattern

A pharmacophoric pattern may be specified by a collection of atoms and distances between them. Heteroatoms and unsaturated carbon atoms will most often be involved in pattern definition, with hydrogens usually not explicitly considered if heteroatoms are labeled as either hydrogen bond donating or accepting. Of the forces involved in drug-receptor binding—coulombic, dipole, hydrogen bond, charge transfer, hydrophobic, and van der Waals, in order of decreasing energy (KOROLKOVAS, 1970)—only the last two are important for saturated carbon. Computer representation of pharmacophoric patterns is discussed in a later section.

A pharmacophoric pattern may be *chiral*, except for a planar pattern, which may be *oriented*. We define pattern chirality in a manner similar

Fig. 1. Definition of pattern
chirality

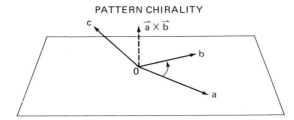

PATTERN CHIRALITY

Fig. 2. Definition of
oriented facial pattern.
z = center of gravity
of the molecule con-
taining the oriented
pattern

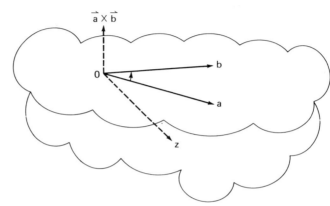

to that used for computer determination of atom chirality (WIPKE and
DYOTT, 1974). For a pattern with its origin at atom 1, construct
vectors $\vec{a}$ and $\vec{b}$ to atoms 2 and 3 (Fig. 1). The cross product $\vec{a} \times \vec{b}$
defines $\vec{a}$ perpendicular direction. Vector $\vec{c}$ drawn to atom 4 then has
the same sign as $\vec{a} \times \vec{b}$ (for parity 0) or opposite sign (for parity 1).
If atom 4 is in the plane of atoms 1-3, then it is replaced by the
first out-of-plane atom for determining chirality. The orienting of
a planar pattern is determined in a similar manner; again, the first
three atoms define a direction $\vec{a} \times \vec{b}$, and that direction should point
away from the center of gravity of a molecule containing an oriented
pattern (Fig. 2). Prochirality of a planar atom is treated similarly
in the re/si nomenclature (HANSON, 1966; GOODWIN, 1973).

A pharmacophoric pattern has the property of *accessibility*. Accessibility may be total, as in cations K^+ and NH_4^+. It may be bifacial,
as in ethidium bromide and other DNA intercalating drugs (WARING,
1970). It may be facial, or facial oriented, as in tetrodotoxin
(MULLINS, 1973). It may be nodal, as in detergents. It may even be
inaccessible; for example, an internal amino acid residue in a globular protein. Operational definitions of the various accessibility
classes are under development. The related concept of atom congestion
has been defined (WIPKE and GUND, 1974). Alternatively, accessibility
may be specified as part of the pattern by defining a negative mass
atom and requiring that the test molecule have no mass at that pattern
position.

Finally, a pattern may be *indefinite*. There may be atom indefiniteness, since often more than one atom type (e.g., NH_2, OH) may be

tolerated at the same position on the receptor surface. This might be specified by indefinite atom types (e.g., D for electron donor atom), or as a fractional electronic charge. An abstraction of this idea defines an interaction pharmacophore as an electrostatic potential field (WEINSTEIN et al., 1973). Also, a functional group (e.g., carboxyl) might be specified as a pattern "superatom," simplifying the number of distance matches to be made. A pattern may also have distance indefiniteness; for example, ZEE-CHENG and CHENG's (1970) antileukemic pattern was reported with distance error limits. The result is a fairly indefinite pattern; as illustrated in Figure 3, the third atom may appear anywhere in the shaded area and still fulfill the defined pattern requirements. Since bonded atoms are relatively close together, they require smaller distance tolerances.

A pattern may also be indefinite because of conformational changes of the drug. For example, the α-adrenergic receptor (KOROLKOVAS, 1970)

Fig. 3. Error limits for antileukemic triangular pattern (ZEE-CHENG and CHENG, 1970)

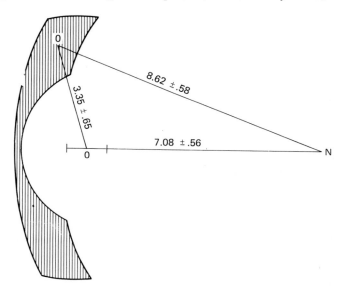

Fig. 4. Proposed α-adrenergic receptor site topology. X = positions where bulk in the drug is not tolerated. From KIER (1968)

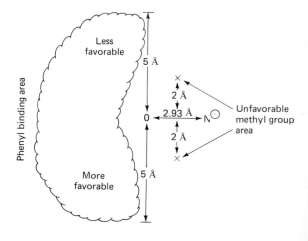

has been formulated with a rather large area for binding of a phenyl ring of active drugs (Fig. 4). This may be accommodated in the pattern by assigning large distance error limits, or rather inelegantly by specifying two pattern distances rather than three. Of course the latter will allow more patterns than desired, but at least all proper patterns will be found. Disallowing conformations with mass at positions marked by X in Fig. 4 would further restrict the number of acceptable patterns. These techniques might also prove useful in cases where the active conformation of a flexible drug was not known.

IV. Three-Dimensional Pharmacophoric Pattern Searching

A. Derivation of Pharmacophoric Patterns

Topologic patterns have typically been proposed as a consequence of structure-activity investigations. Topographic patterns have been derived from mapping of a topologic pattern onto a rigid drug analog, or from determining the preferred conformation of a flexible molecule by theoretical or experimental means. Confidence is gained in the reality of such a pattern by finding it in several different drug structures exhibiting similar bioactivity. A long-term objective is to develop an automatic pattern recognizer that finds patterns common to several drugs possessing the same mode of action. Our initial aim, however, was more modest: to search for presupposed patterns to determine their worth. Some patterns already proposed in the literature are collected in Table 1; where available, references are to reviews.

B. Derivation of Drug Molecule Structures

Molecular structures are available from modeling and from experimental studies. About 5000 molecular structures are available from the Cambridge Crystallographic Data File (ALLEN et al., 1973). Transforming unit cell coordinates to cartesian coordinates is trivial (computer program available from the author). The crystallographic method suffers from limitations, however, in that it is a relatively long and expensive procedure performed by experts on specialized equipment; the crystal structure may correspond neither to the solution conformation nor the bioactive form; and crystal structures are not available for many drugs, all liquids and amorphous powders, and nearly all hypothesized new drugs. Structural information is available from a variety of other experimental methods, including electron diffraction, proton and ^{13}C nmr, and lanthanide shift experiments (BARRY et al., 1974).

Molecular coordinates may be derived from measurements on a Dreiding model by using various computer programs that create molecules from standard fragments (such as programs no. 94, 130, 135, 136, 169, 178, 186, and 226 from Quantum Chemistry Program Exchange, Indiana University, Bloomington, Ind.), or by using computer graphics programs that create a structure from an input two-dimensional structural diagram

Table 1. Some proposed pharmacophoric and receptor patterns

Bioactivity	Pattern	References
Local anesthetic	receptor map	KOROLKOVAS, 1970
Muscarinic	receptor map	KOROLKOVAS, 1970;
	pharmacophore	KIER, 1971; KIER, 1973;
Acetylcholinesterase	receptor map	KOROLKOVAS, 1970
Analgesic	receptor map	KOROLKOVAS, 1970; CASY, 1973
		PORTOGHESE and WILLIAMS, 1969
		TAKEMORI, 1974
Anti-inflammatory	receptor map	KOROLKOVAS, 1970; KIER, 1971
Nicotinic	two-point pattern	KOROLKOVAS, 1970; KIER, 1971;
		KIER, 1973
Anticholinergic	receptor map	KOROLKOVAS, 1970;
	pharmacophore	MAAYANI et al., 1973;
		WEINSTEIN et al., 1973;
		PAULING, 1975
α-,β-Adrenergic	receptor map	KOROLKOVAS, 1970; KIER, 1971;
		KIER, 1973
	pharmacophore	PULLMAN et al., 1972;
		COUBEILS et al., 1972;
		PATIL et al., 1975
Histaminic	receptor map	KOROLKOVAS, 1970
	pharmacophore	KIER, 1971; KIER, 1973
Antihistaminic	receptor surface	
	complement	CROXATTO and HUIDOBRO, 1956
Serotoninic	pharmacophore	KOROLKOVAS, 1970; KIER, 1971;
		KIER, 1973; KELLY and
		ADAMSON, 1973
MAO Inhibitory	receptor map	KOROLKOVAS, 1970
Neuroleptic	topological	JANSSEN, 1973
	pharmacophore	KOROLKOVAS, 1970;
		FEINBERG and SNYDER, 1975
Hallucinogenic	pharmacophore	KANG et al., 1973;
		GREEN et al., 1973
Convulsant	receptor model	SMYTHIES, 1974
Anticonvulsant	pharmacophore	CAMERMAN and CAMERMAN, 1970
Steroid hormonal	receptor maps	KOROLKOVAS, 1970; KIER, 1971;
		KIER, 1973;
		CRENSHAW et al., 1974
Taste	receptor map	BEETS, 1973; GUILD, 1972;
		KIER, 1973
Antileukemic	triangle	ZEE-CHENG and CHENG, 1970;
		ZEE-CHENG and CHENG, 1973;
		ZEE-CHENG et al., 1974
Antipeptic	pharmacophore	BUSTARD and MARTIN, 1972
Hypertensive,	receptor surface	
hypotensive	complements	CROXATTO and HUIDOBRO, 1956
Antimalarial	triangle	CHENG, 1974

(WIPKE, 1974; STILL and LEWIS, 1974; BOLT BERANEK and NEWMAN, 1973;
ZANDER and JURS, 1975; DIERDORF and KOWALSKI, 1974). Computerized
molecular modeling systems are becoming commonplace in universities
(review: MARSHALL et al., 1974), government laboratories (FELDMANN et

al., 1973), and industry [HODGES and NORDBY (Searle), 1975; GUND, ANDOSE and RHODES (Merck, Sharp and Dohme), unpublished].

Several theoretical methods are available for predicting molecular structure (GOLEBIEWSKY and PARCZEWSKI, 1974). In particular, molecular mechanics calculations (WILLIAMS et al., 1968; ENGLER et al., 1973; HOPFINGER, 1973), quantum mechanics methods (POPLE and BEVERIDGE, 1970; KIER, 1970, 1971, 1973; GREEN et al., 1974), and combinations (ALLINGER and SPRAGUE, 1973; WARSHEL and KARPLUS, 1974) have been used. Such calculations, by "conformation mapping," can indicate which structures are most likely to occur (MARSHALL et al., 1974). The magnitude of the current research effort devoted to determining three-dimensional structure of biologically active molecules is apparent from the Proceedings of two recent symposia on the subject (BERGMANN and PULLMAN, 1973, 1974).

The flexibility of many drugs is a worrisome complication. When several conformational states are energetically accessible, then different conformations may prevail in vacuo (the state treated by most theoretical calculations), in solution, in the crystal, and on the receptor surface (GREEN et al., 1974). In fact, it appears that many effector substances (agonists) act by undergoing one or more conformational changes on the receptor surface, while rigid substances are more likely to be antagonists. Indeed, BURGEN et al. (1975) have proposed the "zipper" model for drug-receptor interaction, in which the drug forms a partial "nucleation" complex with the receptor, followed by conformational changes in both drug and receptor to give stronger binding. This model permits fast and selective binding of substrates, and rapid conformational transformation of the biomolecule—which may be required to induce a bioaction.

What, then, is the point of examining molecular models of the non-receptor bound conformation? Basically, because there must be some pattern recognition of the uncomplexed molecule, or there would be no driving force for complexation. BUSTARD and MARTIN (1972) have argued that since according to collision theory the rate of complexation varies with the concentration of the active species, optimal activity will occur for a drug that reacts in its ground state conformation. KIER (1973) has proposed the concept of "remote recognition of preferred conformation," and KIER and HOLTJE (1975) have reported calculations that support the notion that there can be long-range attractive interaction with a drug's ground state conformation.

Our short term solution to the drug flexibility problem is to collect all energetically accessible conformations and pattern search on each one separately. GREEN et al. (1974) recommend considering all conformations within 6 kcal/mole of the global minimum. Alternatively, partially defined patterns (see above) could be used to minimize the importance of a specific conformation. In yet another approach, CHENEY (1974) and SUNDARAM (1975) have used optimization methods to constrain a flexible molecule to superpose upon a more rigid molecule.

C. Finding a Pattern Match

Matching patterns is analogous to the well-known chemical information problem of matching a molecular substructure (fragment). A practical substructure search system generally matches on screens (previously defined fragments or other chemical information) to reduce the problem to a manageable size, then searches for an atom-by-atom match on the reduced set of structures (LYNCH et al., 1971). Similarly, we may save some computer time by counting atom types to assure that the pattern may occur in the present molecule; more sophisticated screening is also possible. We finally need to perform an atom-by-atom match, comparing distances rather than bonds as we go.

Since there are $n(n-1)$ interatomic distances in a molecule of n atoms, any reduction in the required number of distance calculations would speed the search. We define the minimum number of distances required for a preliminary match as $4(n-3)+2$ for patterns of 4 atoms or more. An informal derivation is given here.

A pattern of two loci is uniquely defined by the simple distance between them (line). A three loci pattern is specified by three distances (triangle), and a four loci pattern by six distances (sides of a tetrahedron). Every additional atom is then uniquely located in space by its distances to the first four atoms, provided that the first four are not all in the same plane. (If they are, the first out-of-plane atom is taken as the fourth basis locus.)

Once a preliminary match is found, it may be checked for closer match. Initial matching occurs with a rather wide distance error limit (typically 1.5 Å, except for bonded atoms). This may be refined by comparing distances to a closer tolerance. Other criteria—e.g., stereochemistry, accessibility—may also be checked at this point.

If a true match is found, then the "goodness of match" should be determined. We define a *similarity index* straightforwardly as the root mean square (r.m.s.) deviation of *all* interatomic distances ($n(n-1)$). Similarly, r.m.s. distance differences have been used as a measure of similarity of isomorphous regions of protein crystal structures (DREUTH et al., 1972; RAO and ROSSMANN, 1973; ROSSMAN et al., 1974; NISHIKAWA and OOI, 1974).

Reducing goodness of pattern match to a single number is convenient, and the similarity index may prove to be a useful measure of steric fit in MR and perhaps PR analyses. However, the twin problems of agonist-antagonist activity (too good a fit may turn an agonist drug into an antagonist) and drug flexibility (a pattern will fit some conformations of a drug better than others) require that this new index be applied with caution.

D. Computer Program MOLPAT

The preceding concepts were implemented in a computer program MOLPAT (GUND et al., 1974). This section should be considered a progress report, since program testing and refinement are incomplete. Patterns were obtained from the literature, or by abstracting the functional atoms from a rigid prototype drug structure. A pattern data file consists of a list of coordinate positions and atom types, and atom-atom combinations for which distances are to be computed. Alternatively, no distances are specified and $4(n-3)+2$ distances are automatically calculated at runtime. Finally, if no coordinate information is given, then distances must appear in the file explicitly. This flexible pattern data format (Fig. 5) allows various descriptions of patterns and partial patterns and has the additional advantage that a pattern may be manipulated and displayed as if it were a molecule.

Drug structures were taken from published crystal determinations or were created from two-dimensional graphically input structural diagrams by a strain minimization program. Molecule file data format (Fig. 6) was similar to pattern data, except bonds and their types are given rather than distances.

In practice, the chemist interactively specifies a pattern and a drug molecule, and the program searches for an occurrence of that pattern in the molecule. If one is found, the occurrence is displayed on a CRT display and match statistics printed. The program is then instructed to search for more occurrences of the pattern in the present molecule, input another molecule, or input another pattern. Flow diagrams outlining the program and pattern search logic are given in Fig. 7 and 8. Although the program is tested and debugged, it cannot be considered to be an effective research tool until a number of

Fig. 5. Pharmacophoric pattern data format

ANTILEUKEMIC PHARMACOPHORE					title
3	3				na nb ptype
—	.92	3.22	0.	O	x, y, z coords.,
0.	0.	0.	0		atom type
7.08	0.	0.	N		
1	2				atoms, dist.
1	3				
2	3				

Fig. 6. Molecule data format

```
MORPHINE, X-RAY STRUCTURE
21    25
−2.606 −1.343  1.352  C
−2.352 −2.803  1.423  C
       :
       :
  1    2    2
  2    3    1
       :
       :
```

Fig. 7. Flow diagram:
Program MOLPAT

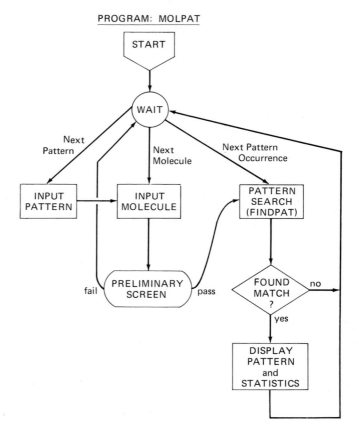

PROGRAM: MOLPAT

program improvements have been implemented—primarily, automatic searching through a file of many drug structures.

The importance of computer graphics for this research should be emphasized (review: MARSHALL et al., 1974). The PDP10/LDS1 facility of the Princeton Computer Graphics Laboratory was utilized for inputting drug structural diagrams and controlling the molecular three-dimensional structure generation, for displaying pharmacophoric patterns in drugs, and for superposing molecules containing similar patterns. While man is the best pattern recognizer, he is best at perceiving two-dimensional patterns. Computer graphics has proven quite helpful in adding a new dimension (the third dimension) to the chemist's perception of structure activity patterns.

V. Pattern Matching Applications

A. Antileukemic Pattern

ZEE-CHENG and CHENG (1970) proposed, from the examination of Dreiding models of drugs, that a triangular arrangement of atoms (Fig. 3) could

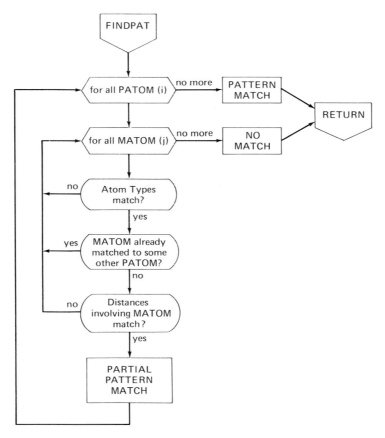

Fig. 8. Flow diagram: Pattern finding subroutine of MOLPAT

be associated with antileukemic activity. They found over 20 anti-leukemic compounds of various chemical classes that contained the pattern and recently have found others (ZEE-CHENG and CHENG, 1973; ZEE-CHENG et al., 1974). We have modeled many of these structures and have confirmed the presence of the hypothesized pattern in most of them. Figure 9 identifies the pattern in a computer-built model of demecolcine.

Fig. 9. Antileukemic triangular pattern located by MOLPAT in a computer-built model of demecolcine

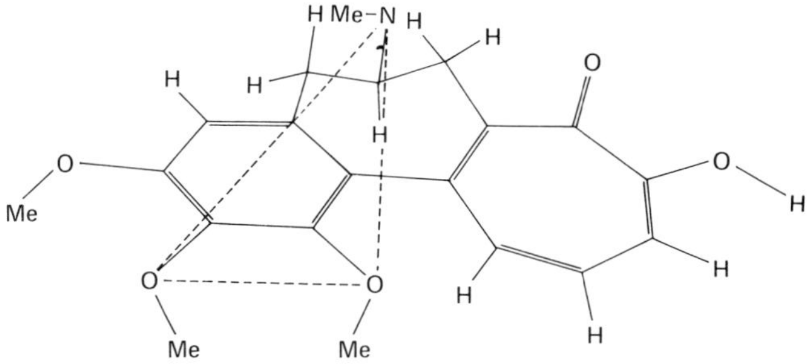

It is certain that these antileukemic agents do not all have the same
mode of action. A subset of these drugs containing a bifacial pattern
(tylocrebrine, tylophorine, streptonigrin, demecolcine) may all act by
intercalation in DNA (ZEE-CHENG and CHENG, 1973). Another subset
consists of known folic acid antagonists (aminopterin, methotrexate).
A subset of nucleoside analogs (sangivamycin, 6-mercaptopurine ribo-
side, cytosine arabinoside, 5-azacytidine) may act either by incorporation in DNA or RNA (5-azacytidine: JUROVCIK et al., 1965) or by
enzyme inhibition (6-mercaptopurine and its anabolites interact with
over 20 enzymes: MONTGOMERY et al., 1970). Other drugs in this group
no doubt have yet other modes of action. It would appear that, in
order to reduce the number of such redundancies, a proposed pharmacophoric pattern should be as detailed as possible.

B. Analgesic Pattern

An analgesic receptor model has been proposed on the basis of observed
drug structure-activity relationships (CASY, 1973; EDDY and MAY, 1973;
SIMON, 1973; CLARKE et al., 1974). Its main features include an aromatic ring and a quaternary N+ atom, separated by an appropriate
distance; a hydroxyl group on the aromatic ring enhances activity, but
is not essential. I modeled this pharmacophoric pattern by extracting
the nitrogen and benzene coordinates from the crystal structure
(MACKAY and HODGKIN, 1955) of the prototypical analgesic, morphine.

The resulting pharmacophoric pattern could be located by MOLPAT in
models of morphine derivatives, in the crystal structure of the morphine antagonist, 3-hydroxy-N-allylmorphinan (BLOUNT et al., 1973),
and, interestingly, in the crystal structure (BAKER et al., 1973) of
LSD (Fig. 10). While LSD is reported to give subjects the illusion

Fig. 10. Analgesic pattern
found by MOLPAT in LSD

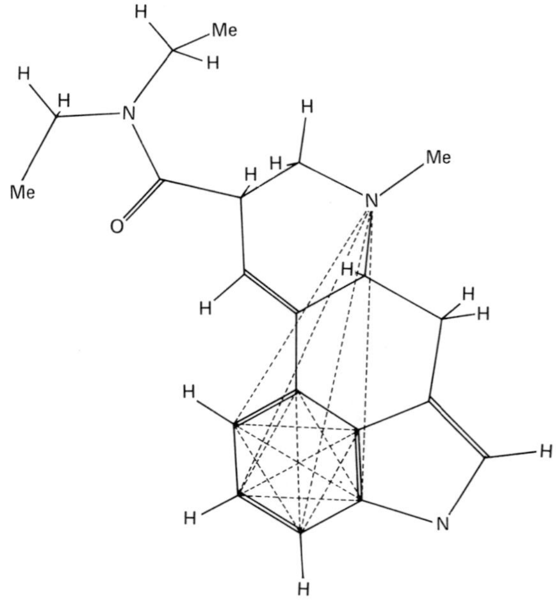

of increased sensory awareness, it actually reduces tactile sensitivity (EDWARDS and COHEN, 1961); furthermore, LSD has been reported to
potentiate response to morphine in a guinea-pig ileum test system
(GINTZLER and MUSACCHIO, 1974).

Recently published binding data of analgesics with an isolated rat
brain receptor (TERENIUS, 1974) may allow refinement and elaboration
of this pharmacophoric pattern, as well as separation of analgesic,
narcotic, and narcotic antagonist activity (TAKEMORI, 1974).

C. Prokaryotic Ribosomal Transpeptidase Inhibitors

While not a direct application of MOLPAT, a recent study (HAHN and
GUND, 1975) illustrates the power of a three-dimensional model for
gaining insight into mechanisms of drug action.

Building on the work of HARRIS and SYMONS (1973), CHENEY (1974) no-
ticed that several antibiotics that inhibit an *in vitro* ribosome
transpeptidase system, as well as the system's natural substrates,
possess similar *topologic* patterns (Fig. 11). SHIPMAN et al. (1974)
used theoretical calculations to find a preferred conformation for
the antibiotic, lincomycin. The ground state was confirmed as the
active form, since the reaction product of lincomycin and formaldehyde—
which has reduced conformational possibilities due to formation of an

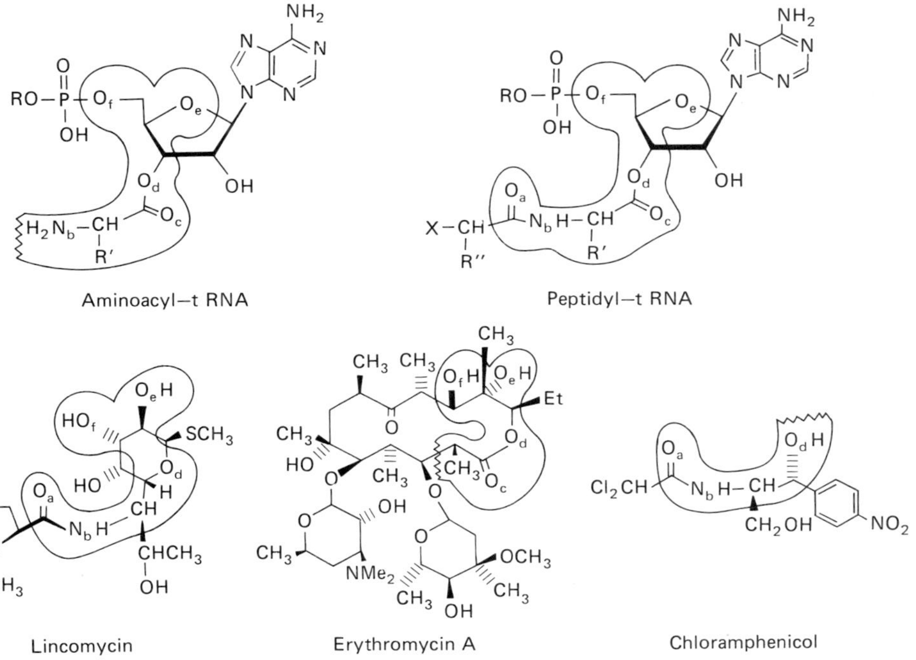

Aminoacyl—t RNA   Peptidyl—t RNA

Lincomycin   Erythromycin A   Chloramphenicol

Fig. 11. Topologic similarities in substrates and some inhibitors of
peptidyl transferase

additional ring—was also antibacterial (SHIPMAN et al., 1974). By
constraining key atoms to their positions in lincomycin and minimizing
differences, CHENEY (1974) derived models for aminoacetyl-tRNA and
peptidyl-tRNA on the enzyme surface. This effectively defined a
topographic pharmacophore for transpeptidase inhibition, if it is
accepted that enzymes enhance reaction rates by stabilizing transition
state geometry of substrates and that enzyme inhibitors may resemble
reactant transition state geometry (LINDQUIST, 1975).

HAHN and GUND (1975) reviewed the large body of information that chemists and biochemists have gathered on chloramphenicol (CM) antibiotic
activity in light of CHENEY's (1974) topographic model of transpeptid-
ation inhibition.

HANSCH analyzed CM analog activity in one of his earliest papers on
multiple regression analysis (HANSCH et al., 1963); subsequently, at
least four other QSAR analyses have been published (CAMMARATA, 1967;
HANSCH et al., 1969, 1973b; HÖLTJE and KIER, 1974). Nevertheless,
KONO et al. (1969) synthesized some analogs that were as active or
more active than CM itself, none of which could have been predicted as
being active from the published equations. Another inadequacy of the
HANSCH correlations was revealed when comparison of *in vivo* and *in
vitro* SAR's (HAHN and GUND, 1975) indicated that steric bulk on the
acetyl group gave excellent binding but poor drugs, presumably due to
poor facilitated transport. Such a separation of varying steric effects upon binding and upon transport is difficult or impossible to
obtain by QSAR techniques. Our three-dimensional structural model
also allowed dismissal of HANSCH's hypothesis that CM acts by forming
a stable free radical (HANSCH et al., 1969). Superposition of chloramphenicol on CHENEY's model for the arrangement of the natural substrates on the enzyme (Fig. 12) results in a model that accomodates
essentially all the SAR data, as well as most of the detailed biochemical data, and that may serve as a framework for the design of
novel antibacterial agents.

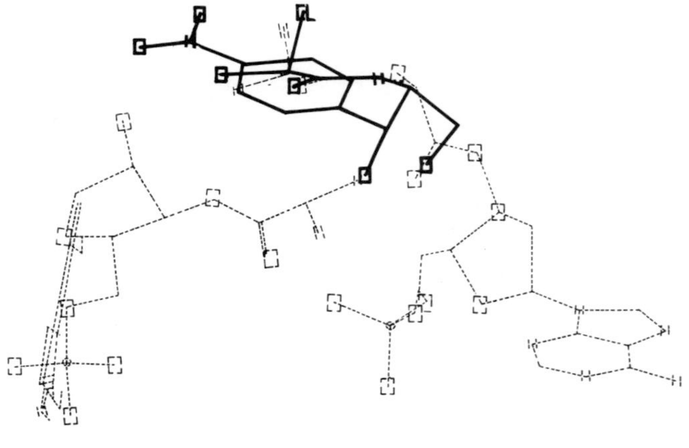

Fig. 12. Model for
chloramphenicol inhi-
bition of transpep-
tidase: Superposition
on model of bound sub-
strates (HAHN and
GUND, 1975)

VI. Conclusions

The drug-receptor theory of biologic action implies a pattern recognition response of receptor to drug, and that response is to a geometrically well-defined (topographic) pattern. Such pharmacophoric patterns may be defined in terms of atom type and interatomic distances, and, as this work has demonstrated, may be manipulated and matched by computer.

Finding patterns in three-dimensional drug structures is a useful method for testing hypothesized pharmacophores, for predicting activity for new compounds, for designing novel drugs with specific activity, and for proposing and testing mechanisms of drug action. Patterns may be used in multiple regression and pattern recognition analyses of drug activity, either qualitatively as present or absent, or quantitatively via the similarity index. Such patterns appear to provide a satisfactory method of accounting for steric effects in biologic reactions.

Acknowledgments

I am grateful to the National Institutes of Health for a Special Research Fellowship (1971-1973), for the period in which most of this research was accomplished. The Princeton Computer Graphics Laboratory is a Biomolecular Resource supported by the National Institutes of Health and Princeton University. I thank Professor W. T. WIPKE for providing invaluable advice and helpful criticism, and Professor R. LANGRIDGE for his interest and support.

References

ADAMSON, G.W., BUSH, J.A.: Method for relating the structure and properties of chemical compounds. Nature (Lond.) 248, 406 (1974).

ALLEN, F.H., KENNARD, O., MOTHERWELL, W.D.S., TOWN, W.G., WATSON, D.G.: Cambridge crystallographic data centre. II. Structural data file. J. Chem. Doc. 13, 119 (1973).

ALLINGER, N.L., SPRAGUE, J.T.: Calculation of the structures of hydrocarbons containing delocalized electronic systems by the molecular mechanics method. J. Am. Chem. Soc. 95, 3893 (1973).

ARIËNS, E.J.: Molecular pharmacology, a basis for drug design. Progr. Drug Res. 10, 429 (1966).

BAKER, R.W., CHOTHIA, C., PAULING, P., WEBER, H.P.: Molecular structures of hallucinogenic substances: lysergic acid diethylamide, psilocybin and 2,4,5-trimethoxyamphetamine. Mol. Pharmacol. 9, 23 (1973). I thank Prof. Pauling for supplying the molecular coordinates.

BARRY, C.D., GLASEL, J.A., WILLIAMS, R.J.P., XAVIER, A.V.: Quantitative determination of conformations of flexible molecules in solution using lanthanide ions as nuclear magnetic resonance probes: application to adenosine-5'-monophosphate. J. Mol. Biol. <u>84</u>, 471 (1974).

BEETS, M.G.J.: Structure-response relationships in chemoreception. In: Structure-Activity Relationships (Ed. C.J. CAVALLITO), vol. I, p. 225. Oxford: Pergamon Press 1973.

BERGMAN, E.D., PULLMAN, B. (Eds.): Conformation of Biological Molecules and Polymers. Proc. Int. Symp., Jerusalem April 1972. Jerusalem: Israel Acad. Sciences and Humanities 1973.

BERGMAN, E.D., PULLMAN, B. (Eds.): Molecular and Quantum Pharmacology. Proc. Int. Symp., Jerusalem April 1974. Dordrecht: Reidel 1974.

BLOUNT, J.F., MOHACSI, E., VANE, F.M., MANNERING, G.J.: Isolation, X-ray analysis and synthesis of a metabolite of (-)-3-hydroxy-N-allylmorphinan. J. Med. Chem. <u>16</u>, 352 (1973).

BOLT BERANEK and NEWMAN, INC.: PROPHET implementation overview. Report No. 2526, April 1973 to Biotechnology Resource Branch. Div. Res. Resources, Nat. Inst. Health.

BREMSER, W., KLIER, M., MEYER, E.: Mutual assignment of subspectra and substructures—a way to structure elucidation by ^{13}C NMR spectroscopy. Org. Magn. Res. <u>7</u>, 97 (1975).

BURGEN, A.S.V., ROBERTS, G.C.K., FEENEY, J.: Binding of flexible ligands to macromolecules. Nature (Lond.) <u>253</u>, 753 (1975).

BUSTARD, T.M., MARTIN, Y.C.: Conformational and structural relationships among antipeptic ulcer compounds. J. Med. Chem. <u>15</u>, 1101 (1972).

CAMERMAN, A., CAMERMAN, N.: Diphenylhydantoin and diazepam: molecular structure similarities and steric basis of anticonvulsant activity. Science <u>168</u>, 1457 (1970).

CAMMARATA, A.: An apparent correlation between the *in vitro* activity of chloramphenicol analogs and electronic polarizability. J. Med. Chem. <u>10</u>, 525 (1967).

CASY, A.F.: Stereochemistry and biological activity. In: Medicinal Chemistry (Ed. A BURGER), 3rd ed., part I, p. 81. New York: Interscience 1970.

CASY, A.F.: Analgesic receptors. In: A Guide to Molecular Pharmacology-Toxicology (Ed. R.M. FEATHERSTONE), part I, p. 217. New York: Dekker 1973.

CAVALLITO, C.J.: Some trends in the development of structure-activity relationships (SAR) and theory of receptors. In: Structure-Activity Relationships (Ed. C.J. CAVALLITO), vol. I, p. 1. Oxford: Pergamon Press 1973.

CHARTON, M.: Steric effects. I. Esterification and acid-catalyzed hydrolysis of esters. J. Am. Chem. Soc. <u>97</u>, 1552 (1975).

CHENEY, V.: *Ab initio* calculations on large molecules using molecular fragments. Structural correlations between natural substrate moieties and some antibiotic inhibitors of peptidyl transferase. J. Med. Chem. <u>17</u>, 590 (1974).

CHENG, C.C.: Novel common structural features among several classes of antimalarial agents. J. Pharm. Sci. <u>63</u>, 307 (1974).

CHU, K.C.: Applications of artificial intelligence to chemistry. Use of pattern recognition and cluster analysis to determine the pharmacological activity of some organic compounds. Anal. Chem. <u>46</u>, 1181 (1974).

CHU, K.C., FELDMANN, R.J., SHAPIRO, M.B., HAZARD, G.F., GERAN, R.I.: Pattern recognition and structure-activity relationship studies.

Computer-assisted prediction of antitumor activity in structurally
diverse drugs in an experimental mouse brain tumor system. J. Med.
Chem. 18, 539 (1975).

CITRI, N.: Conformational adaptability in enzymes. Adv. Enzymol. 37,
397 (1973).

CLARKE, R.L., GAMBINO, A.J., DAUM, S.J.: In pursuit of analgetic
agents. Hydro-1,3-ethanoindeno-[2,1-c]pyridines and homologs. J.
Med. Chem. 17, 1040 (1974).

CLERC. J.T., NAEGELI, P., SEIBL, J.: Artificial intelligence. Chimia
27, 639 (1973).

COUBEILS, J.L., COURRIÈRE, Ph., PULLMAN, B.: Quantum-mechanical
study of the conformational properties of sympatholytic compounds.
J. Med. Chem. 15, 453 (1972).

CRAMER, R.D., III, REDL, G., BERKOFF, C.E.: Substructural analysis. A
novel approach to the problem of drug design. J. Med. Chem. 17, 533
(1974).

CRENSHAW, R.R., LUKE, G.M., JENKS, T.A., BIALY, G.: Potential anti-
fertility agents. 7. Synthesis and biological activities of 2-, 3-,
and 6-alkyl-substituted 4-aryl-2-methylcyclohexanecarboxylic acids.
J. Med. Chem. 17, 1262 (1974).

CROXATTO, R., HUIDOBRO, F.: Fundamental basis of the specificity of
pressor and depressor amines in their vascular effects. Arch. Int.
Pharmacodyn. 106, 207 (1956).

DETAR, DeL. F.: Calculation of steric effects in reactions. J. Am.
Chem. Soc. 96, 1254 (1974a).

DETAR, DeL. F.: Quantitative predictions of steric acceleration. J.
Am. Chem. Soc. 96, 1255 (1974b).

DIERDORF, D.S., KOWALSKI, B.R.: Three-dimensional molecular structure-
biological activity correlation by pattern recognition. U.S.N.T.I.S.,
Ad Rep. No. 785863/2GA (1974).

DRENTH, J., HOL, W.G.J., JANSONIUS, J.N., KOEKOEK, R.: A comparison of
the three-dimensional structures of subtilisin BPN' and subtilisin
novo. In: Structure and Function of Proteins at the Three-Dimensional
Level. Cold Spring Harbor Symp. Quant. Biol. 36, p. 107. Cold Spring
Harbor Laboratories 1972.

DUBOIS, J-E.: DARC system in chemistry. In: Computer Representation
and Manipulation of Chemical Information (Eds. W.T. WIPKE, S.R.
HELLER, R.J. FELDMANN, E. HYDE), p. 239. New York: Interscience 1974.

DUBOIS, J-E, LAURENT, D., ARANDA, A.: Systéme DARC. XVII. Théorie de
topologie-information. II. Méthode PELCO-procédure d'établissement
de corrélation de topologie-information. J. Chim. Physique 70, 1616
(1973).

EAKIN, D.R., HYDE, E.: Evaluation of on-line techniques in a substructure search system. In: Computer Representation and Manipulation
of Chemical Information (Eds. W.T. WIPKE, S.R. HELLER, R.J. FELDMANN,
E. HYDE), p. 1. New York: Interscience 1974.

EDDY, N.B., MAY, E.L.: The search for a better analgesic. Science 181,
407 (1973).

EDWARDS, A.E., COHEN, S.: Visual illusion, tactile sensibility and
reaction time under LSD-25. Psychopharmacologia 2, 297 (1961).

EHRLICH, P.: Über den jetzigen Stand der Chemotherapie. Chem. Ber. 42,
17 (1909).

ENGLER, E.M., ANDOSE, J.D., SCHLEYER, P.v.R.: Critical evaluation of
molecular mechanics. J. Am. Chem. Soc. 95, 8005 (1973).

FEINBERG, A.P., SNYDER, S.H.: Phenothiazine drugs: structure-activity relationships explained by a conformation that mimics dopamine. Proc. Nat. Acad. Sci. 72, 1899 (1975).

FELDMANN, R.J.: Interactive graphic chemical structure searching. In: Computer Representation and Manipulation of Chemical Information (Eds. W.T. WIPKE, S.R. HELLER, R.J. FELDMANN, E. HYDE), p. 55. New York: Interscience 1974.

FELDMANN, R.J., BACON, C.R.T., COHEN, J.S.: Versatile interactive graphics display system for molecular modelling by computer. Nature (Lond.) 244, 113 (1973).

FENNER, H.: EPR studies on the mechanism of biotransformation of tricyclic neuroleptics and antidepressants. In: The Phenothiazines and Structurally Related Drugs (Eds. I.S. FORREST, C.J. CARR, E. USDIN), p. 5. New York: Raven 1974.

FREE, S.M., Jr., WILSON, J.W.: A mathematical contribution to structure-activity studies. J. Med. Chem. 7, 395 (1964).

FUJITA, T., IWASA, J., HANSCH, C.: A new substituent constant, π, derived from partition coefficients. J. Am. Chem. Soc. 86, 5175 (1964).

GINTZLER, A.R., MUSACCHIO, J.M.: Interaction between serotonin and morphine in the guinea-pig ileum. J. Pharmacol. Exp. Ther. 189, 484 (1974).

GOLEBIEWSKI, A., PARCZEWSKI, A.: Theoretical conformational analysis of organic molecules. Chem. Rev. 74, 519 (1974).

GOODFORD, P.J.: Prediction of pharmacological activity by the method of physicochemical-activity relationships. Adv. Pharmacol. Chemother. 11, 51 (1973).

GOODWIN, T.W.: Prochirality in biochemistry. In: Essays in Biochemistry (Eds. P.N. CAMPBELL, F. DICKENS), vol. IX, p. 103. London: Academic Press 1973.

GOURLEY, D.R.H.: Biological responses to drugs. In: Medicinal Chemistry (Ed. A. BURGER), 3rd ed., part I, p. 25. New York: Interscience 1970.

GRAY, C.J.: Enzyme-Catalyzed Reactions, p. 86. London: van Nostrand 1971.

GREEN, J.P., DRESSLER, K.P., KHAZAN, N.: Mescaline-like activity of 2-amino-7-hydroxytetralin. Life Sci. 12, part I, 475 (1973).

GREEN, J.P., JOHNSON, C.L., KANG, S.: Application of quantum chemistry to drugs and their interactions. Ann. Rev. Pharmacol. 14, 319 (1974).

GUILD, W., Jr.: Theory of sweet taste. J. Chem. Educ. 49, 171 (1972).

GUND, P., WIPKE, W.T., LANGRIDGE, R.: Computer searching of a molecular structure file for pharmacophoric patterns. In: Proc. Int. Conf. Computers in Chem. Res. and Educ., Ljubljana, July 12-17, 1973, vol. 3, p. 5/33. Amsterdam: Elsevier 1974.

HAHN, F.E., GUND, P.: A structural model of the chloramphenicol receptor site. In: Topics in Infectious Diseases, Vol. I Drug Receptor Interactions in Antimicrobial Chemotherapy (Eds. J. DREWS, F.E. HAHN), p. 245. Vienna: Springer 1975.

HAMMETT, L.P.: The effect of structure upon the reactions of organic compounds. Benzene derivatives. J. Am. Chem. Soc. 59, 96 (1937).

HANSCH, C.: A quantitative approach to biochemical structure-activity relationships. Accounts Chem. Res. 2, 232 (1969).

HANSCH, C.: Quantitative approaches to pharmacological structure-activity relationships. In: Structure-Activity Relationships (Ed. C.J. CAVALLITO), vol. I, p. 75. Oxford: Pergamon Press 1973.

HANSCH, C., KUTTER, E., LEO, A.: Homolytic constants in the correlation of chloramphenicol structure with activity. J. Med. Chem. $\underline{12}$, 746 (1969).

HANSCH, C., LEO, A., UNGER, S.H., KIM, K.H., NIKAITANI, D,, LIEN, E.J.: "Aromatic" substituent constants for structure-activity correlations. J. Med. Chem. $\underline{16}$, 1207 (1973a).

HANSCH, C., MALONEY, P.P., FUJITA, T., MUIR, R.M.: Correlation of biological activity of phenoxyacetic acids with Hammett substituent constants and partition coefficients. Nature (Lond.) $\underline{194}$, 178 (1962).

HANSCH, C., MUlR, R.M., FUJITA, T., MALONEY, P.P., GEIGER, F., STREICH, M.: The correlation of biological activity of plant growth regulators and chloromycetin derivatives with Hammett constants and partition coefficients. J. Am. Chem. Soc. $\underline{85}$, 2817 (1963).

HANSCH, C., NAKAMOTO, K., GORIN, M., DENISEVICH, P., GARRETT, E.R., HEMAN-ACKAH, S.M., WON, C.H.: Structure-activity relationships of chloramphenicols. J. Med. Chem. $\underline{16}$, 917 (1973b).

HANSON, K.R.: Applications of the sequence rule. I. Naming the paired ligands g,g at a tetrahedral atom Xggij. II. Naming the two faces of a trigonal atom Yghi. J. Am. Chem. Soc. $\underline{88}$, 2731 (1966).

HANSON, K.R., ROSE, I.A.: Interpretations of enzyme reaction stereospecificity. Accounts Chem. Res. $\underline{8}$, 1 (1975).

HARRIS, R.J., SYMONS, R.H.: On the molecular mechanism of action of certain substrates and inhibitors of ribosomal peptidyl transferase. Bioorganic Chem. $\underline{2}$, 266 (1973).

HILLER, S.A., GOLENDER, V.E., ROSENBLIT, A.B., RASTRIGIN, L.A., GLAZ, A.B.: Cybernetic methods of drug design. I. Statement of the problem—the perceptron approach. Computers Biomed. Res. $\underline{6}$, 411 (1973).

HODGES, D., NORDBY, D.H.: MOLOCH-3: the G. D. Searle molecular modeling system. Abstr. COMP-7, 169th Am. Chem. Soc. Nat. Meeting, Phila., April 1975.

HÖLTJE, H-D., KIER, L.B.: A theoretical approach to structure-activity relationships of chloramphenicol and congeners. J. Med. Chem. $\underline{17}$, 814 (1974).

HOPFINGER, A.J.: Conformational Properties of Macromolecules. New York: Academic Press 1973.

JANSSEN, P.A.J.: Chemical and pharmacological classification of neuroleptics. In: The Neuroleptics (Eds. D.P. BOBON, P.A.J. JANSSEN, J. BOBON), p.33. Basel: Karger 1970.

JANSSEN, P.A.J.: Structure-activity relationships (SAR) and drug design as illustrated with neuroleptic agents. In: Structure-Activity Relationships (Ed. C.J. CAVALLITO), vol. I, p. 37. Oxford: Pergamon Press 1973.

JOHNSON, C.D.: The Hammett Equation. Cambridge: University Press 1973.

JURS, P.C., ISENHOUR, T.L.: Chemical Applications of Pattern Recognition. New York: Interscience 1975.

KANG, S., JOHNSON, C.L., GREEN, J.P.: Theoretical studies on the conformation of psilocin and mescaline. Mol. Pharmacol. $\underline{9}$, 640 (1973).

KAUFMAN, J.J., KERMAN, E.: Quantum chemical and other theoretical techniques for the understanding of the psychoactive action of phenothiazines. In: The Phenothiazines and Structurally Related Drugs (Eds. I.S. FORREST, C.J. CARR, E. USDIN), p. 55. New York: Raven 1974.

KAUFMAN, J.J., KOSKI, W.S.: Physicochemical, quantum chemical and other theoretical techniques for the understanding of the mechanism of action of CNS agents. Psychoactive drugs, narcotics, and narcotic antagonists and anesthetics. In: Drug Design (Ed. E.J. ARIENS), vol. 5, p. 251. New York: Academic Press 1975.

KELLY, J.M., ADAMSON, R.H.: A comparison of common interatomic distances in serotonin and some hallucinogenic drugs. Pharmacology 10, 28 (1973).

KENT, P., GAÜMANN, T.: Considerations on the interpretation of mass spectra via learning machines. Helv. Chem. Acta 58, 787 (1975).

KIER, L.B.: The preferred conformations of ephedrine isomers and the nature of the *alpha* adrenergic receptor. J. Pharmacol. Exp. Ther. 164, 75 (1968).

KIER, L.B.: Receptor mapping using molecular orbital theory. In: Fundamental Concepts in Drug-Receptor Interactions (Eds. J.F. DANIELLI, J.F. MORAN, D.J. TRIGGLE), p. 15. New York: Academic Press 1970.

KIER, L.B.: Molecular Orbital Theory in Drug Research. New York: Academic Press 1971.

KIER, L.B.: The prediction of molecular conformation as a biologically significant property. Pure Appl. Chem. 35, 509 (1973).

KIER, L.B., HÖLTJE, H-D.: A stochastic model of the remote recognition of preferred conformation in a drug-receptor interaction. J. Theoret. Biol. 49, 401 (1975).

KONO, M., O'HARA, K., HONDA, M., MITSUHASHI, S.: Drug resistance of staphylococci. XI. Induction of chloramphenicol resistance by its derivatives and analogues. J. Antibiot. (Tokyo) 22, 603 (1969).

KOROLKOVAS, A.: Essentials of Molecular Pharmacology, Background for Drug Design. New York: Interscience 1970.

KOWALSKI, B.R.: Pattern recognition in chemical research. In: Computers in Chemical and Biochemical Research (Eds. C.E. KLOPFENSTEIN, C.L. WILKINS), vol. 2, p. 1. New York: Academic Press 1974.

KOWALSKI, B.R., BENDER, C.F.: The application of pattern recognition to screening prospective anticancer drugs. Adenocarcinoma 755 biological activity test. J. Am. Chem. Soc. 96, 916 (1974).

KOWALSKI, B.R., BENDER, C.F.: Solving chemical problems with pattern recognition. Naturwissenschaften 62, 10 (1975).

LEFFLER, J.E., GRUNWALD, E.: Rates and Equilibria of Organic Reactions. New York: Wiley 1963.

LIEN, E.J.: The use of substituent constants and regression analysis in the study of structure-activity relationships. Am. J. Pharm. Educ. 33, 368 (1969).

LINDQUIST, R.N.: The design of enzyme inhibitors: transition state analogs. In: Drug Design (Ed. E.J. ARIENS), vol. 5, p. 24. New York: Academic Press 1975.

LYNCH, M.F.: The microstructure of chemical data-bases and the choice of representation for retrieval. In: Computer Representation and Manipulation of Chemical Information (Eds. W.T. WIPKE, S.R. HELLER, R.J. FELDMANN, E. HYDE), p. 31. New York: Interscience 1974.

LYNCH, M.F., HARRISON, J.M., TOWN, W.G., ASH, J.E.: Computer Handling of Chemical Structural Information. London: MacDonald 1971.

MAAYANI, S., WEINSTEIN, H., COHEN, S., SOKOLOVSKY, M.: Acetylcholine-like molecular arrangement in psychomimetic anticholinergic drugs. Proc. Nat. Acad. Sci. 70, 3103 (1973).

MACKAY, M., HODGKIN, D.C.: A crystallographic examination of the structure of morphine. J. Chem. Soc. 3261 (1955).

MARSHALL, G.R., BOSSHARD, H.E., ELLIS, R.A.: Computer modeling of chemical structures: applications in crystallography, conformational analysis, and drug design. In: Computer Representation and Manipulation of Chemical Information (Eds. W.T. WIPKE, S.R. HELLER, R.J. FELDMANN, E. HYDE), p. 203. New York: Interscience 1974.

MATHEWS, R.J.: A comment on structure-activity correlations obtained using pattern recognition methods. J. Am. Chem. Soc. 97, 935 (1975).

McMAHON, R.E.: Drug metabolism. In: Medicinal Chemistry (Ed. A. BURGER), 3rd ed., part I, p. 50. New York: Interscience 1970.

MONTGOMERY, J.A., JOHNSTON, T.P., SHEALY, Y.F.: Drugs for neoplastic diseases. In: Medicinal Chemistry (Ed. A. BURGER), 3rd ed., part I, p. 680. New York: Interscience 1970.

MULLINS, L.J.: The use of models of the cell membrane in determining the mechanism of drug action. In: A Guide to Molecular Pharmacology-Toxicology (Ed. R.M. FEATHERSTONE), part I, p. 1. New York: Dekker 1973. See esp. p. 34.

NISHIKAWA, K., OOI, T.: Comparison of homologous tertiary structures of proteins. J. Theoret. Biol. 43, 351 (1974).

NORRINGTON, F.E., HYDE, R.M., WILLIAMS, S.G., WOOTTEN, R.: Physico-chemical-activity relationships in practice. I. A rational and self-consistent data bank. J. Med. Chem. 18, 604 (1975).

OLDENDORF, W.H.: Blood-brain barrier permeability to drugs. Ann. Rev. Pharmacol. 14, 239 (1974).

ÖTVÖS, L., MORAVCSIK, E., KRAICSOVITS, F.: Stereochemistry of the reactions of biopolymers. V. Steric effect of chiral substituents in enzyme-catalyzed reactions. Tetrahed. Letters 2485 (1975).

PATIL, P.N., MILLER, D.D., TRENDELENBURG, U.: Molecular geometry and adrenergic drug activity. Pharmacol. Rev. 26, 323 (1975).

PAULING, P.: The shapes of cholinergic molecules. In: Cholinergic Mechanisms (Ed. P.G. WASER), p. 241. New York: Raven 1975.

PERRIN, C.L.: Testing of computer-assisted methods for classification of pharmacological activity. Science 183, 551 (1974).

POPLE, J.A., BEVERIDGE, D.L.: Approximate Molecular Orbital Theory. New York: McGraw-Hill 1970.

PORTOGHESE, P.S., WILLIAMS, D.A.: Stereochemical studies on medicinal agents. VII. Absolute stereochemistry of methadol isomers and the role of the 6-methyl group in analgetic activity. J. Med. Chem. 12, 839 (1969).

PULLMAN, B., COUBEILS, J-L., COURRIÈRE, Ph., GERVOIS, J-P.: Quantum mechanical study of the conformational properties of phenethyl-amines of biochemical and medicinal interest. J. Med. Chem. 15, 17 (1972).

PURCELL, W.P., BASS, G.E., CLAYTON, J.M.: Strategy of Drug Design: A Molecular Guide to Biological Activity. New York: Interscience 1973.

RAŠKA, K., Jr., JUROVČÍK, M., ŠORMOVÁ, Z., ŠORM, F.: Inhibition by 5-Azacytidine of Ribonucleic Acid Synthesis in Isolated Nuclei of Calf Thymus. Collec. Czech. Chem. Commun. 30, 3215 (1965).

RAO, S.T., ROSSMANN, M.G.: Comparison of super-secondary structures in proteins. J. Mol. Biol. 76, 241 (1973).

REDL, G., CRAMER, R.D., III, BERKOFF, C.E.: Quantitative drug design. Chem. Soc. Rev. 3, 273 (1974).

ROSSMANN, M.G., MORAS, D., OLSEN, K.W.: Chemical and biological evolution of a nucleotide-binding protein. Nature (Lond.) 250, 194 (1974).

SEGAL, H.L.: Enzymatic interconversion of active and inactive forms of enzymes. Science 180, 25 (1973).

SHIPMAN, L.L., CHRISTOFFERSEN, R.E., CHENEY, B.V.: *Ab initio* calculations on large molecules using molecular fragments. Lincomycin studies. J. Med. Chem. 17, 583 (1974).

SHORTER, J.: The separation of polar, steric, and resonance effects by the use of linear free energy relationships. In: Advances in Linear Free Energy Relationships (Eds. N.B. CHAPMAN, J. SHORTER), p. 71. London: Plenum Press 1972.

SHORTER, J.: Correlation Analysis in Organic Chemistry: an Introduction to Linear Free-Energy Relationships. Oxford: Clarendon Press 1973.

SIMON, E.J.: In search of the opiate receptor. Am. J. Med. Sci. 266, 160 (1973).

SIMON, Z: Specific interactions. Intermolecular forces, steric requirements, and molecular size. Ang. Chem. Int. Ed. 13, 719 (1974).

SIMON, Z., SZABADAI, Z.: Minimal steric difference parameter and the importance of steric fit for structure-biological activity correlations. Studia Biophys. (Berlin) 39, 123 (1973).

SINKULA, A.A., YALKOWSKY, S.H.: Rationale for design of biologically reversible drug derivatives: prodrugs. J. Pharm. Sci. 64, 181 (1975).

SMYTHIES, J.R.: Relationships between the chemical structure and biological activity of convulsants. Ann. Rev. Pharmacol. 14, 9 (1974).

STILL, W.C., LEWIS, A.J.: Computerized molecular modeling. Chemtech. 4, 118 (1974).

STUPER, A.J., JURS, P.C.: Classification of psychotropic drugs as sedatives or tranquilizers using pattern recognition techniques. J. Am. Chem. Soc. 97, 182 (1975).

SUNDARAM, K.: Applications of an optimization technique in submolecular biology. Int. J. Quantum Chem. 9, 393 (1975).

SWAIN, C.G., LUPTON, E.C.: Field and resonance components of substituent effects. J. Am. Chem. Soc. 90, 4328 (1968).

TAFT, R.W., LEWIS, I.C.: Evaluation of resonance effects on reactivity by application of the linear inductive energy relationship. V. Concerning a σ_R scale of resonance effects. J. Am. Chem. Soc. 81, 5343 (1959).

TAKEMORI, A.E.: Biochemistry of drug dependence. Ann. Rev. Biochem. 43, 15 (1974).

TERENIUS, L.: Contribution of "receptor" affinity to analgesic potency. J. Pharm. Pharmacol. 26, 146 (1974).

TING, K-L.H., LEE, R.C.T., MILNE, G.W.A., SHAPIRO, M., GUARINO, A.M.: Applications of artificial intelligence: relationships between mass spectra and pharmacological activity of drugs. Science 180, 417 (1973).

WARING, M.: Binding of drugs to supercoiled circular DNA: evidence for and against intercalation. Progr. Mol. Subcell. Biol. 2, 216 (1970).

WARSHEL, A., KARPLUS, M.: Calculation of $\pi\pi^*$ excited state conformations and vibronic structure of retinal and related molecules. J. Am. Chem. Soc. 96, 5677 (1974).

WEINSTEIN, H., MAAYANI, S., SREBRENIK, S., COHEN, S., SOKOLOVSKY, M.: Psychotomimetic drugs as anticholinergic agents. II. Quantum-mechanical study of molecular interaction potentials of

1-cyclohexylpiperidine derivatives with the cholinergic receptor. Mol. Pharmacol. 9, 820 (1973).

WELLS, P.R.: Linear Free Energy Relationships. London: Academic Press 1968.

WILLBRANDT, W., ROSENBERG, T.: The concept of carrier transport and its corollaries in pharmacology. Pharmacol. Rev. 13, 109 (1961).

WILLIAMS, J.E., STANG, P.J., SCHLEYER, P.v.R.: Physical organic chemistry: quantitative conformational analysis; calculation methods. Ann. Rev. Phys. Chem. 19, 531 (1968).

WIPKE, W.T.: Computer-assisted three-dimensional synthetic analysis. In: Computer Representation and Manipulation of Chemical Information (Eds. W.T. WIPKE, S.R. HELLER, R.J. FELDMANN, E. HYDE), p. 147. New York: Interscience 1974.

WIPKE, W.T., DYOTT, T.M.: Simulation and evaluation of chemical synthesis. Computer representation and manipulation of stereochemistry. J. Am. Chem. Soc. 96, 4825 (1974).

WIPKE, W.T., GUND, P.: Congestion: a conformation-dependent measure of steric environment. Derivation and application in stereoselective addition to unsaturated carbon. J. Am. Chem. Soc. 96, 299 (1974).

ZANDER, G.S., JURS, P.C.: Generation of mass spectra using pattern recognition techniques. Anal. Chem. 47, 1562 (1975).

ZEE-CHENG, K.Y., CHENG, C.C.: Common receptor-complement feature among some antileukemic compounds. J. Pharm. Sci. 59, 1630 (1970).

ZEE-CHENG, K.Y., CHENG, C.C.: Interaction between DNA and coralyne acetosulfate, an antileukemic compound. J. Pharm. Sci. 62, 1572 (1973).

ZEE-CHENG, K.Y., PAULL, K.D., CHENG, C.C.: Experimental antileukemic agents. Coralyne, analogs, and related compounds. J. Med. Chem. 17, 347 (1974).

Chemical Evolution: A Terrestrial Reassessment

Norman W. Gabel

I. Introduction

A. Do Chemicals Evolve?

There is at the time of this writing a great profusion of books, articles, and experimental papers dealing with subjects pertaining to chemical evolution and the origin of life. One more progress report encompassing all of this work would serve little purpose at this time. Therefore, no endeavor has been made to make this report a comprehensive treatise. Instead, an examination of some of the basic assumptions of chemical evolution is attempted. For those readers for whom the subject matter is unfamiliar, a balanced and unbiased introductory account can be found in RUTTEN's (1971) *The Origin of Life by Natural Causes*. Detailed accounts from several different points of view are listed in a supplementary bibliography.

The term "chemical evolution" was suggested by CALVIN (1956, 1959, 1961) to describe the process by which geochemical events that preceded biological evolution could have led to the origin of life. Several other expressions such as molecular evolution and prebiotic chemistry are often used to denote this process or the investigations that are directed toward it. A problem that some scientists have had with this terminology is that molecules in themselves do not evolve in any manner; however, it must be remembered that individual biological entities also do not evolve but merely change. It is the process of emergence of species in a hierarchical manner that is generally considered to be biological evolution. In the same sense, a molecule is an individual entity composed of two or more atoms that are covalently bonded. A chemical, on the other hand, refers to a distinct population of molecules, atoms, or ions (WHELAND, 1949). It is often argued, when philosophizing upon science, that the terminology of one field cannot be explicitly applied to another field of investigation. This is a truism, but that type of argument is surely an exercise in hair-splitting, since no two people ever mean exactly the same thing when using the same word. In this context, the process by which populations of chemicals abiotically give rise to other chemicals of an emerging hierarchy during the course of geological history does seem most fittingly called "chemical evolution."

B. Historical Perspective

Current investigations in chemical evolution had their inception in the hypothesis set forth by OPARIN (1924, 1938, 1964) and HALDANE (1928). OPARIN, as a chemist, speculated that, in a terrestrial environment with little or no molecular oxygen, the other gases of the atmosphere would be subjected to the photolyzing radiation of the sun and would give rise to a complex mixture of organic chemicals that could be used as the framework for contemporary living systems. HALDANE came to a very similar conclusion after he became convinced that anaerobic organisms were phylogenetically more primitive than aerobic organisms. The physicist BERNAL (1951) strongly supported and promulgated these views. RUTTEN (1971) points out historically that all three men were influenced by Marxist doctrine in their goal of a completely materialistic theory of life, its evolution, and origin.

The simulated laboratory investigation of chemical evolution received its single most powerful stimulus from the successful experiments of MILLER (1953, 1955). Adopting UREY's (1952) assumption that the primitive atmosphere of the earth was highly reducing, MILLER (1953) subjected a mixture of methane, ammonia, and water to electric discharges as a simulation for lightning. A large number of biologically relevant organic compounds were formed including amino acids and carboxylic acids. The publication of these results was followed by a literal avalanche of experiments by other investigators (cf. PATTEE, 1965; PONNAMPERUMA and GABEL, 1968). Biologically relevant organic compounds were found to be easily synthesized by subjecting this highly reduced system of gases to various sources of energy. Within any given series of homologous compounds, the molecules of the lowest molecular weight were present in the greatest abundance. Branching of the carbon chain and the identification of many compounds not found in the contemporary biosphere was characteristic of these abiotic random syntheses.

Based upon the success of these experiments and the fact that the universe (RUSSELL, 1935) and the solar system (BROWN, 1952) are chemically dominated by hydrogen, MILLER and UREY (1959), by considering the rate at which hydrogen escapes from the earth today, calculated a possible pressure of hydrogen on the surface of the earth 4.5 billion years ago to be 1.5×10^{-3} atmospheres. The pressure of methane under these conditions was estimated to be 4×10^3 atmospheres. To account for these highly reducing conditions, HOLLAND (1962) presented a model for the primitive atmosphere in which he added the suggestion that, in the early history of the earth, iron metal was still present in large quantities in the crust and upper mantle. If the majority of the atmosphere was ejected from primordial volcanoes under these conditions, the gases would have been highly reducing and would have consisted of H_2O, CO, CO_2, CH_4, N_2, and H_2. At 1200°C, the methane/carbon monoxide ratio would have been 1.2 and, upon cooling, virtually all of the carbon molecules would have been converted to methane.

In contrast, it should be noted that RUBEY (1955) and ABELSON (1956, 1966) were never convinced that the primordial atmosphere was highly reducing. RUBEY contended that hydrogen would escape much too rapidly from the primitive earth to sustain an atmosphere dominated by methane and nitrogen. ABELSON pointed out that a methane-dominated atmosphere should have left behind, in the geological record, wide-spread evidence of abiotically produced hydrocarbons. Experiments by ABELSON (1966) and his associates demonstrated that a primordial atmosphere, resulting from the same gases that emanate from contemporary volcanoes, could also produce biologically relevant organic chemicals.

C. Popular Conceptions

The popular conception of terrestrial chemical evolution and the origin of life is based on the assumption that the gases emitted via primordial volcanism were highly reducing and must have resulted in an atmosphere that contained methane, ammonia, and water as its principal constituents. Various sources of energy, such as ultraviolet light from the sun, heat from volcanoes, electric discharges in the form of lightning, and ionizing radiation from radionuclides, acting upon this reducing atmosphere, produced a large quantity and number of organic compounds until the early oceans had the consistency of a hot dilute soup. The two main gas phase products that have been detected in these reactions are hydrogen cyanide and formaldehyde. These two compounds are considered to be intermediates in the abiotic primordial synthesis of many biologically relevant organic monomers, and hydrogen cyanide and its oligomers are thought to have acted as aqueous-phase condensing agents in the formation of peptides and nucleosides. Phosphorylation of the nucleosides then produced nucleotides. Self-replicating systems of molecules eventually appeared which, through their interactions, gave rise to contemporary carbon-based life. This sequence of events is believed to have been duplicated in innumerable planetary systems throughout the universe (SHKLOVSKII and SAGAN, 1966).

As outgassing eventually became less reducing and carbon dioxide accumulated in the atmosphere, the photochemically produced organic chemicals became depleted in the hot dilute soup. Those organisms (autotrophs) that could photosynthetically convert carbon dioxide to organic substrates and molecular oxygen continued to flourish, while heterotrophic organisms survived only if their particular environment was replenished with organic substrates by autotrophic organisms. During this transition from a reducing to an oxidizing atmosphere, it has been postulated that there was a wholesale biologic massacre (PONNAMPERUMA and GABEL, 1968). The organisms that survived and developed further evolved into the wide variety of life on earth today (Fig. 1).

II. Development of the Terrestrial Environment

A. The Reality of Pilot-Plant Chemistry

When highly reducing gases are incrementally mixed with an oxidizing gas, the resulting mixtures are explosive throughout a broad range

Fig. 1. Transition from reducing
to oxidizing atmosphere

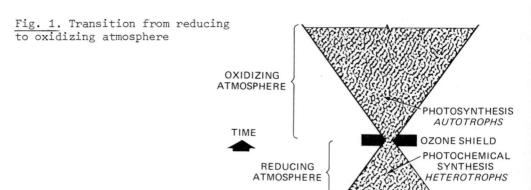

of composition.[1] Pilot-plant studies of vapor-phase reactions involving hydrocarbons and a gaseous oxidizing agent are designed to be
conducted outside the explosion limits of the reacting mixture of
gases. The explosion limits for any particular mixture are derived
from empirical observations determined on a small scale. The data
that have been compiled from these investigations have not, heretofore,
been applied to the development of the terrestrial atmosphere as depicted in Fig. 1. The descriptive scenario of the preceding section
assumes that the composition of highly reduced gases being emitted by
primordial volcanoes would eventually change to a more oxidized mixture in which carbon dioxide was the most abundant carbon compound.
This mixture would be similar in composition to the fumes of contemporary volcanic outgassing.

Inasmuch as ethane is not flammable with carbon dioxide in any portion
when less than 13.4% molecular oxygen is present (LEWIS and VON ELBE,
1951a), it can be assumed that the atmospheric hydrocarbons could have
gradually reacted with CO_2 to form carbon monoxide and water as the
principal products:

$$3CO_2 + CH_4 \longrightarrow 4CO + 2H_2O$$

Any molecular oxygen that would have been produced by photolysis or
photosynthesis would also react gradually with the volatile hydrocarbons to produce aldehydes that via decarbonylation, could also increase the carbon monoxide concentration of the atmosphere:

$$CH_4 + O_2 \longrightarrow CH_2O + H_2O$$

$$CH_2O \longrightarrow CO + H_2$$

Based on the kinetics of the reaction of CO and OH to form formate
ion in alkaline solutions, VAN TRUMP and MILLER (1973) calculated that

$$CO + OH^- \rightleftharpoons HCOO^-$$

---

[1]The fundamental importance of this basic fact of gas dynamics, as it
applies to chemical evolution, was pointed out to me by Dr. E.J.
GRIFFITH, Monsanto Company, St. Louis, Missouri.

the half-life of CO in the primordial atmosphere would have been
12×10^6 years at $0°C$ and 5.5×10^4 years at $25°C$. Even though VAN
TRUMP and MILLER proposed that the ocean acted as an efficient sink
for primordial CO and prevented it from becoming a significant atmospheric constituent, it is difficult, in the absence of any evidence
for copious prebiologic deposits of organic matter, to envision how
4×10^3 atmospheres of methane could be converted to biomass and CO_2
without CO becoming at least a minor constituent of a $CO-CO_2-N_2$ atmosphere during the transition from a highly reducing mixture to a
highly oxidizing mixture of gases.

The maximum nonexplosive percentage of molecular oxygen in carbon
monoxide at any measured concentration that is diluted with CO_2 or N_2
is 5.9 and 5.6% of the CO concentration, respectively (LEWIS and VON
ELBE, 1951b). As photosynthesis continued, the molecular oxygen would
have eventually exceeded this maximum percentage in the $CO-CO_2-N_2$
atmosphere and all increasingly oxygen-rich mixtures would have been
subject to violent explosions when exposed to lightning or vulcanism,
until the level of oxygen exceeded its level of flammability with
carbon monoxide.

During the last 3 billion years of the earth's geological history,
there is no evidence for the occurrence of an expansive thermal catas-
trophe (RUTTEN, 1971). A carbon monoxide explosion could have been
avoided only if carbon monoxide were never more than a trace constituent of the earth's atmosphere. If this is the case, then it is highly
unlikely that the earth's atmosphere ever went through a transition
from a highly reducing mixture of gases to a strongly oxidizing mixture of gases.

B. Actualism vs. Catastrophism

Geological thought is at present based on the principle of actualism,
i.e., processes now operative on and in the earth are utilized to
explain the events of geologic history (RUTTEN, 1971). (RUTTEN pre-
fers the word "actualism" to the nearly equivalent "uniformitarianism.")
Actualism replaced the earlier principle of catastrophism, which was
the popular concept at the beginning of the last century. Catastrophism postulated catastrophies of great magnitude occurring at specific
times in geological history. These catastrophies led to faunal crises,
mass extinctions, etc., and were presumed to be due to causes not at
present operative in the contemporary environment.

Actualism does not deny that sudden catastrophies have occurred, and
still occur, but that they are never world-wide and their effect is
never of comparable magnitude to processes that are actually taking
place at a slow rate over a long time span. Furthermore, variations
in these processes have always been of a quantitative nature rather
than a qualitative one.

RUTTEN (1962) previously had termed the anoxygenic period of the early
Precambrian as "preactualistic." However, in his last book RUTTEN
(1971) extended the actualistic philosophy of geology to this early

anoxygenic environment and then went further to accommodate the chemists' desire for a highly reducing environment. It would appear now that the latter accommodation of geology to prebiotic terrestrial chemistry was unwarranted.

In the arguments set forth by RUBEY (1955) and ABELSON (1966), the source of the primordial atmosphere was the continual outgassing of volcanoes that continues into our present age. The effluent of contemporary volcanoes consists primarily of H_2O, CO_2, and N_2. Small quantities of H_2, CO, HCl, H_2S, and SO_2 are also present. ABELSON's (1966) description of the formation of the hydrosphere and the atmosphere is based on actualism and is probably as close to what actually happened as anyone will ever get. The acidic gases being emitted by the volcanoes were neutralized through the weathering of silicate minerals, producing an ocean with inorganic components similar to what is present today. The buffering system of weathered silicates and the equilibrium between atmospheric and oceanic CO_2 would tend to maintain the pH at 8-9. Although most of the CO would be converted to formate ion in the ocean (BRANCH, 1915), a trace amount would be left in the atmosphere. The reduced valence state of some early Precambrian sedimentary ores may have been due to local concentrations of formate ion.

The two principal components of the primordial atmosphere were most likely N_2 and CO_2. From geological observations, the concentration of CO_2 was never greater than 10 times its present level of 0.03% (RUTTEN, 1971). What the trace components were and what role they played in atmospheric reactions is something that can only be ascertained after extensive meteorological modeling and considerable experimentation.

Recent developments in the study of the solar system (CAMERON, 1973a,b) have led WALKER (1976) to propose an inhomogeneous accretion model for the formation of the earth. The primarily carbonaceous late-forming veneer would have had an overall oxidation state that was approximately the same as that of the upper layers of the earth today. WALKER presents a convincing argument that the primordial atmosphere might have resembled the modern atmosphere with the addition of a few percent of hydrogen and the removal of all oxygen.

C. The Prebiotic Formation of Organic Compounds

1. Atmospheric Reactions

If the assumption is made that the organic matter in the crust of the primitive earth volatilized during outgassing, then prebiotic organic molecules were formed from CO_2, N_2, and several trace constituents within the atmosphere and from water, carbonate, formate, and probably NH_3 in the hydrosphere. Unlike the literature on the production of organic compounds in a reducing atmosphere, there is a paucity of information on vapor phase reactions in a predominantly N_2-CO_2 atmosphere. During the production of organic compounds from CO_2, the oxidation number of carbon must decrease. There are only two ways by which such an event will occur: (1) dissociation of CO_2 to CO and atomic oxygen, and (2) chemical reduction. Unless a rather appreciable quantity of atomic or molecular oxygen escaped during the early formation of the atmosphere, the former process [even though it is chemically feasible (BARTH, 1970; WIEGAND and NIGHAN, 1973)] would probably make an insignificant contribution to the primordial accumulation of organic molecules.

Insofar as outgassing was responsible for the accumulation of prebiotic organic molecules, the extent of their formation would have been dependent on the reduced components being emitted by volcanoes. ABELSON (1966) estimates that the outgassed CO could have produced a formate concentration in the primordial ocean of 0.0035 M. If all of the reducing capacity of outgassed hydrogen were employed to produce CO, then the oceanic concentration of formate would have been as high as 0.6 M. Formate concentrations of this magnitude probably never existed, but the figures are indicative of the amount of reduced volatiles available for organic synthesis.

In the absence of H_2, the most significant photochemical product of trace quantities of CO in an N_2 atmosphere is C_3O_2 (HARTECK et al., 1938; LIUTI et al., 1969):

$$4CO \longrightarrow C_3O_2 + CO_2$$

Carbon suboxide is a very reactive gas and liquid; bp 7°C at 761 mm Hg; d = 1.1 (0°C). It polymerizes or reacts with any nucleophile spontaneously. In the presence of water, it is immediately converted to malonic acid:

$$O=C=C=C=O + 2H_2O \longrightarrow HOOC-CH_2-COOH$$

A review that discusses the polymerization of carbon suboxide and its reactions with organic nucleophiles to form heterocyclic compounds has been published (BUKOWSKI and POREJKO, 1969).

PERLS (1971) has examined the constraints on the quantity of C_3O_2 polymer that could be produced in the Martian atmosphere and has speculated that various forms of C_3O_2 polymer may be responsible for the color changes on Mars. RAFF and MEABURN (1969) had previously proposed a hypothetical role for the reactions of C_3O_2 and CO in the primitive atmosphere of earth based on the assumption of continual CO outgassing. HUBBARD et al. (1971) subjected CO and water vapor to ultraviolet photolysis in an attempt to simulate the Martian atmosphere and claimed to have detected the formation of formaldehyde, acetaldehyde, and glycolic acid.

Although active nitrogen (electronically excited N or N_2) does not appear to be reactive with CO or CO_2 (WRIGHT and WINKLER, 1968; SIMONAITIS and HEICKLEN, 1973), YOUNG and MORROW (1974) recently reported the detection of CN radical as a direct product of CO and active nitrogen. In ABELSON's (1956, 1966) original experiments on a hypothetical atmosphere of volcanic gases, HCN was produced as a major product from ultraviolet photolysis of N_2-CO-H_2 mixtures, but

the partial-pressure ratios of the gases were 8:8:2 proceeding upward
to richer mixtures of CO and H_2. These concentrations of CO and H_2
are probably much too high for any sustained primordial atmosphere.
Even under these conditions, nitriles were not detected, and in the
absence of hydrogen, HCN was not produced. Abundant quantities of HCN
and its oligomers may not have existed in the primordial environment.
Unquestionably, some HCN was synthesized from the small amounts of H_2
and CO in the N_2-CO_2 atmosphere; and the reaction of active nitrogen
with malonic acid or acetic acid (via decarboxylation of malonic acid)
in the vapor phase could also have produced some HCN. Nonetheless,
the relevance of HCN with regard to prebiotic chemistry should be re-
examined after its quantitative determination in experiments conducted
with an anoxygenic, primarily N_2-CO_2 atmosphere.

As further evidence that the role of HCN and aliphatic nitriles may
have been grossly overrated in previous investigations of prebiotic
chemistry, FERRIS and CHEN (1975) reported that, even if the primitive
atmosphere were composed of N_2, CH_4, and H_2O, ultraviolet photolysis
of this mixture produces alcohols, aldehydes, ketones, and questionable traces of hydrocarbons. Nitriles and HCN were not detected.

2. Hydrospheric Reactions

The two carbon compounds that were probably present in the greatest
concentrations in the primordial ocean were carbonate and formate ions.
Formic acid is sometimes used as a reducing agent in the laboratory
synthesis of organic compounds. In the absence of oxygen, GARRISON et
al. (1951) reported the reduction of carbon dioxide in aqueous solutions to formic acid and formaldehyde by ionizing radiation. High-
energy helium-ion or proton irradiation of dilute aqueous formic acid
solutions at neutral or slightly basic pH led to the synthesis of a
number of products of higher molecular weight (GARRISON et al., 1952,
1958). The compounds that were isolated and identified included
oxalic, glyoxylic, glycolic, mesoxalic, tartronic, and tartaric acids.
Glyoxal and formaldehyde were isolated and identified as their 2,4-
dinitrophenylhydrazones.

$$HOOCCOOH$$
oxalic acid

$$\underset{HCCOOH}{\overset{O}{\overset{\|}{}}}$$
glyoxylic acid

$$HOCH_2COOH$$
glycolic acid

$$\underset{OH}{\overset{OH}{HOOC-C-COOH}}$$
mesoxalic acid

$$\underset{}{\overset{OH}{HOOC-CH-COOH}}$$
tartronic acid

$$\underset{}{\overset{OH\ OH}{HOOC-CH-CH-COOH}}$$
tartaric acid

$$\underset{HC-CH}{\overset{O\ O}{\overset{\|\ \|}{}}}$$
glyoxal

GETOFF et al. (1960) found that, when aqueous CO_2 was subjected to γ irradiation from a ^{60}Co source in the absence of oxygen, formaldehyde and acetaldehyde were produced in acid, neutral, and basic solutions. At neutrality, the concentration of the two products continued to increase throughout the course of irradiation. A dose of 10^{18} eV produced 10^{15} molecules of aldehyde. Formic acid, oxalic acid, and hydrogen, as well as many other unidentified compounds, were formed in smaller amounts. Formaldehyde was also easily produced from CO_2 in a $FeSO_4$ solution by irradiation at 2537 Å with a low-pressure mercury vapor lamp. The authors concluded that aqueous CO_2 could be reduced by any radiation process involving hydrogen atoms and solvated electrons.

In the primordial environment, both acetaldehyde and malonic acid could be expected to give rise to acetic acid. GARRISON et al. (1953) irradiated acetic acid (0.25 M) with cyclotron-produced helium ions with an energy of 35 MeV. At doses less than 10^{20} eV per ml, succinic acid was the main product. At higher doses, succinic, tricarballylic, malonic, malic, and citric acids were identified by isolation on silicic acid columns. Many unidentified compounds were also present. These results were substantiated by TANAKA and WANG (1967) during an investigation of the radiolysis of succinic acid in water with a ^{60}Co source. In addition to the aforementioned compounds, they isolated and reported oxalacetic acid. NEGRON-MENDOZA and PONNAMPERUMA (1976) have also irradiated aqueous acetic acid for the purpose of studying chemical evolution and have found essentially the same results, but were also able to identify many of the minor components through combined gas chromatography and mass spectrometry.

```
CH2-COOH              CH2-COOH              COOH
|                     |                     |
CH2-COOH              CH -COOH              CH2
                      |                     |
                      CH2-COOH              COOH
succinic acid
                      tricarballylic acid   malonic acid


COOH                  CH2-COOH              COOH
|                     |                     |
CHOH                  HO-C-COOH             C=O
|                     |                     |
CH2                   CH2-COOH              CH2
|                                           |
COOH                  citric acid           COOH

malic acid                                  oxalacetic acid
```

The point of the foregoing discussion was to demonstrate that a great variety of organic acids could have been produced in the primordial ocean from CO_2 and formic acid via photolysis at the surface and radiolysis by β and γ rays from radionuclides as dissolved cations. As an example, the decay of ^{40}K in the crust and surface of the earth today gives rise to 3 x 10^{19} calories per year; 2.6 x 10^9 years ago, this would have been 12 x 10^{19} calories per year (SWALLOW, 1960). In addition, the energy released by cavitation processes (ANBAR, 1968) also may have played an effective role.

In the absence of effective quantities of cyanide ion, amino acids could still have been synthesized easily from the carboxylic acids. HARADA and IWASAKI (1974) have shown that amino acids are produced during the glow-discharge electrolysis of carboxylic acids dissolved in aqueous ammonia.[2] When ammonia is subjected to a glow discharge, the effective reactant is active nitrogen (WRIGHT and WINKLER, 1968). An electric discharge, such as lightning, passing through an N_2 atmosphere and being grounded by a body of water that contained carboxylic acids, should also be expected to produce amino acids. The resulting mixture of amino acids will probably more clearly resemble the distributional pattern of amino acids in biological material than the distribution produced by the Miller-Urey synthesis. (Glutamic and aspartic acids are at present more abundant in nature than the other amino acids.) The degradation of these synthesized amino acids could have served to introduce small amounts of cyanide, urea, cyanate, and other proposed primordial monomers to the environment.

It should be emphasized that, during the geological time-span in which it is postulated that organic compounds were synthesized from carbonate, formate, N_2, and water, the rate and amount of organic material that would have been synthesized were ultimately dependent upon the reducing capacity of the gaseous mixture emitted by volcanoes. The appearance of the photosynthetic organism overcame this potentially limiting factor by using the energy of the sun to reduce CO_2 with water. If photosynthetic organisms had not eventually appeared on this planet, the entire biomass of the earth could have been limited by the amount of CO and H_2 released through vulcanism. Photosynthesis was an evolutionary giant step.

3. Weathering of the Lithosphere

In the inhomogeneous accretion model for the formation of the earth, the late-forming veneer of carbonaceous chondrites was probably not subjected to evenly distributed heating and concomitant volatilization of organic material. Some contribution to the organic molecules of the hydrosphere may have come from the weathering of meteoritically derived material. Specifically, some thermally stable heterocyclic pigments, such as porphyrins, may have had meteorites as their source.

The carbonaceous chondrites are considered to be a primitive condensate from the solar nebula, and they contain a rich variety of complex organic compounds (ANDERS et al., 1974; LAWLESS et al., 1972). These organic compounds are believed to have formed under reducing conditions beyond the planetary orbit of Mars. Hydrocarbons, carboxylic acids, heterocyclic nitrogenous bases, and amino acids have been extracted and identified from the Murchison, Murray, and Mighei meteorites (KVENVOLDEN et al., 1970; CRONIN and MOORE, 1971; PERING and

---

[2]The significance of the experiment by HARADA and IWASAKI was brought to my attention by Dr. Patricia BUHL, Department of Chemistry, University of Maryland, College Park, Maryland.

PONNAMPERUMA, 1971; LAWLESS et al., 1971; ANDERS et al., 1973; BUHL,
1975). Within the capabilities of the analytical scheme, the asymmetric compounds that were extracted were present as racemic mixtures.
HODGSON and BAKER (1969) have detected pigments resembling porphyrins
in these meteorites, but it is not clear whether these were actually
porphyrins or linear polymers of pyrroles.

There is some question as to whether Miller-Urey or Fischer-Tropsch
reaction conditions prevailed during the formation of the carbonaceous
chondrites (ANDERS et al., 1974). Either set of reaction conditions
can account for most of the organic compounds that have been extracted
from the meteorites. One outstanding difference, however, is that
Fischer-Tropsch reaction conditions (CO, H_2, NH_3, or N_2, and a metal
oxide catalyst) give an isotopic fractionation of $^{12}C/^{13}C$ that is
similar in magnitude and sign to meteoritic organic matter (LANCET and
ANDERS, 1970) whereas the Miller-Urey reaction conditions are reported
to result in virtually no isotopic fractionation (LANCET, 1972).

The bulk of the organic material (70-90%) of carbonaceous chondrites
is in the form of high molecular weight, nonextractable material.
BITZ and NAGY (1966, 1967), upon subjecting the nonextractable organic
material from the Orgueil meteorite to ozonolysis, found that the
resulting aromatic acids were very similar to those obtained from the
ozonolysis of bituminous coal. This nonextractable organic matter
would not seem to be a material that could be easily weathered from
the lithosphere of the primordial earth. The one possibly significant
contribution of organic molecules leached directly from carbonaceous
meteoritic material into the earth's prebiotic chemical milieu may
have been some of the structurally complex, aromatic, heterocyclic
molecules that are functionally important in biologic oxidation-
reduction reactions.

III. Further Chemical Evolutionary Hierarchies

A. Polymerization

As a generalization, structural and functional systems of living
things at the molecular level are constructed from repeating units of
biomonomers such as amino acids, monosaccharides, nucleosides, polyhydroxy alcohols, carboxylic acids, and phosphates. (Even the mineral
matrix of vertebrate bones has a three-dimensional repeating sequence.)
Assuming a sufficient quantity of biomonomers in the primordial environment, the next major hierarchical step in chemical evolution would
be the formation of biopolymers. The formation of a biopolymer inevi-
tably involves a dehydration-type of condensation reaction. In aqueous solutions, dehydration reactions are very improbable spontaneous
events.

FOX (1974a,b) and his associates had proposed in the 1950's that biomonomers were thermally dehydrated to biopolymers. Specifically,
mixtures of amino acids, when in contact with freshly erupted volcanic

material, produced polymeric materials that were called "proteinoids" and which, upon rehydration, formed microspherules (FOX and HARADA, 1958). These microspheres, sometimes thermally produced in admixture with nucleotides, have even been proposed as prototypes of the primitive cell (FOX et al., 1971a,b). Although some thermally dehydrated amino acid polymers can be synthesized under milder conditions than vulcanism, it is difficult to envision in the scheme described by FOX, how the "correct" proportions of amino acids necessary to produce proteinoids accumulated in specific locations on the primordial earth and then withstood thermal decomposition after being exposed to volcanic temperatures. Analyses of contemporary volcanic material disclose the presence of only insignificant amounts of amino acids (Table 1). Furthermore, the amino acids that were detected occur commonly in biologic material and could be the result of contamination during collection of the samples. An additional argument against the prebiotic relevance of thermally produced amino acid polymers is that the polylysine produced in this manner is a branched, cross-linked polymer (HEINRICH et al., 1969). Lysine is one of the important main ingredients used in the preparation of basic proteinoids.

Many simple derivatives of HCN have been found to promote dehydration-type condensations in aqueous solution, e.g., cyanamide, dicyanamide, dicyandiamide, cyanogen, and HCN tetramer. MILLER and ORGEL (1973) have questioned the prebiologic significance of these reactions because the HCN derivatives react more readily with the solvent molecules, water, than with the biomonomers. HULSHOF and PONNAMPERUMA (1975) recently reviewed the prebiotic relevance of primordial condensing agents and concluded that linear polyphosphates or cyclic metaphosphates are the most plausible prebiotic condensing agents. Only HCN tetramer and the condensed phosphates are effective in neutral to slightly basic solutions and, as MILLER and ORGEL (1973)

Table 1. Amino acid analysis of samples collected from recent eruptions on the volcanic island Heimaey off the southern coast of Iceland[a]

| | Large crater[b] | | Small crater[b] | | Lava flow[c] | | Procedural blank | |
	H_2O[d]	HCl[e]	H_2O	HCl	H_2O	HCl	H_2O	HCl
Glycine[f]	6	24	3	13	–	12	–	1
Alanine	3	8	3	17	–	5	–	1/2

[a]Samples were collected and analyzed during spring of 1973 by Dr. Akira SHIMOYAMA, Laboratory of Chemical Evolution, Dept. of Chemistry, University of Maryland, College Park, Maryland 20742.
[b]Volcanic ash was collected several meters inside rim of volcano 1-7 days after cessation of an eruption.
[c]Lava was collected from red, free-running flow.
[d]Extracted with water at 100°C for 24 h.
[e]Extracted with 6N HCl at 100°C for 24 h.
[f]Quantities of amino acid are expressed as ng/g of sample. Traces of aspartic acid, glutamic acid, threonine, and serine were also detected. Samples were analyzed with a Durrum Automated Amino Acid Analyzer.

indicated, the concentration of HCN tetramer required for a dehydration-type condensation reaction may have been prohibitively high.

HOROWITZ (1945) postulated 30 years ago that the first protolife entity used for its structural network and energy metabolism those prefabricated molecules that were initially present in the environment. Inorganic polyphosphates and metaphosphates could have been produced by several different prebiotic mechanisms (GRIFFITH et al., 1976), and they have been shown to be widely distributed within the contemporary biosphere (HAROLD, 1966; GABEL and THOMAS, 1969, 1971; GLONEK et al., 1971) in a polymodal functional capacity (DEIERKAUF and BOOIJ, 1968; KULAEV, 1971; GABEL, 1972). Two naturally occurring igneous minerals have been reported that have a condensed phosphate composition within their crystalline structures (SEMENOV et al., 1962; BELOV and ORGANOVA, 1962). These two minerals, metalomonosovite and lomonosovite, were found to occur in pegmatite granites of the Lovozero Massif adjacent to the Khibiny Massif on the Kola Peninsula, U.S.S.R. VINOGRADOV and TUGARINOV (1961) calculated that the igneous rocks of the Kola Peninsula solidified 3.6×10^9 years ago. Insofar as there is validity to the postulate of HOROWITZ, these two early Precambrian minerals could be considered as evidence for the role of inorganic poly- and metaphosphates as primordial dehydration-type condensing agents.

It was suggested 10 years ago (GABEL, 1965) that inorganic polyphosphates were continually leached from primordial igneous rocks and fed into the streams and runoff water flowing toward the shoreline of large bodies of water. The brackish shoreline and estuarine waters would have had a continual supply of inorganic condensed phosphates during this weathering process. Although it is doubtful that the concentration of phosphorus in the general expanses of the seas was ever greater than it is now, crustal differentiation and weathering would have ultimately led to localized shoreline environments with appreciable quantities of soluble orthophosphate and condensed phosphates (GRIFFITH et al., 1976). It is a misconception that polyphosphates are insoluble in the presence of calcium under all conditions of basic pH. They are soluble in a seawater or brackish environment as long as the molar concentration of the phosphate approaches the concentration of the calcium (GABEL, 1965; GRIFFITH, 1972). (The aqueous solubility of calcium complexes of the condensed phosphates is one of their most industrially important physicochemical properties.) In brackish primordial waters, organic molecules of the immediate environment that could be electrostatically attracted or that could coordinate with the exterior metal cations of polyphosphate complexes, would form a film or envelope around this macromolecular structure. Through this "scavenging" effect, the organic molecules would be brought into contact with the condensing agent, and the condensation reactions would not be subject to the kinetic control of random collisions in a dilute solution (GABEL, 1965). Considering the postulated paucity of organic material in primordial waters, this "scavenging" effect of poly- and metaphosphates may be one of their most efficacious properties in prebiotic chemistry.

RABINOWITZ et al. (1969) were the first investigators to successfully demonstrate that peptides were produced in dilute aqueous solutions of

amino acids and inorganic polyphosphates at moderate temperatures and
neutral to slightly basic conditions. The percent conversions varied
considerably in succeeding studies (RABINOWITZ, 1970, 1972; CHUNG et
al., 1971; LOHRMANN and ORGEL, 1973), and, although a mechanistic
study was given by CHUNG et al. (1971), these authors concluded that
the reaction had no prebiologic significance, since they did not believe that a source of condensed phosphates would have been available
in the primordial environment. A critical examination of the mechanism of this reaction is given by HULSHOF and PONNAMPERUMA (1975).
LOHRMANN and ORGEL (1973) later changed their minds about the prebiologic significance of condensed phosphates.

One factor that may account for the large discrepancy in reported
yields of oligopeptides produced by this reaction is the amount and
kind of cations in the aqueous medium. Although RABINOWITZ and his
associates reported that their experiments were performed in distilled
water, most commercial and laboratory sources of distilled water contain variable trace amounts of mono- and divalent cations. HULSHOF
(unpublished data, 1974) observed that, when highly purified sodium
trimetaphosphate and glycine were subjected to the conditions described by CHUNG et al. (1971) in water that had been purified for
geochemical analyses, no peptides could be detected in the reaction
medium. When tap water was substituted for the purified water, the
reaction proceeded remarkably well. Obviously, this reaction requires
at least trace quantities of some ubiquitous cation, probably Ca^{2+} or
Mg^{2+}, in order to proceed. This same type of phenomenon is demonstrated by numerous biologic phosphohydrolases (ALBERS, 1967; GABEL, 1972).

Currently, an attempt is being made to determine whether a linear
inorganic polyphosphate could have served as template, reactant, and
condensing agent for the formation of a prebiotic oligonucleotide
(GABEL, 1976). Calcium cation apparently serves preferentially as an
external counterion to the helical coil of linear inorganic polyphosphates in water (GLONEK et al., 1975). Ribose, the sugar monomer of
RNA, forms a much stronger complex with calcium cation than any other
pentose (MILLS, 1961). Inasmuch as pentoses are the major product of
the clay-catalyzed oligomerization of dilute solutions of formaldehyde
(GABEL and PONNAMPERUMA, 1967), inorganic polyphosphates, in the primordial environment, could have preferentially selected calcium as an
external counterion and the calcium, in turn, may have selectively
complexed with ribose or perhaps even a preformed ribonucleoside.
This macromolecular complex may have been responsible, via a concerted
alcoholysis of the phosphoric anhydride bonds by the ribose hydroxyl
groups to form sugar-phosphate ester-bonds, for the primordial appearance of the first oligonucleotide.

In preliminary experiments, not only has the preference of calcium
cation for ribose been corroborated but it was found that the base
cytosine and the nucleosides cytidine and deoxycytidine also migrated
as calcium complexes. This association with calcium was not observed
with any other bases or nucleosides. The relative stabilities of the
complexes of the sugars, bases, and nucleosides with calcium was
determined by their electrophoretic mobility in a calcium acetate
electrolyte at pH 8. Electrophoretic mobilities of this type have

been demonstrated to be useful for determining the relative stabilities of complexes of alkali and alkaline-earth metals (RENDLEMAN,
1966). Whether a macromolecular complex of the type involving inorganic polyphosphate, calcium cations, and nucleosides can result in
an oligonucleotide, remains to be demonstrated.

The only other condensed phosphate that has been used in the preparation of peptides or oligonucleotides is the mixture of phosphate
ethyl esters obtained from ethyl ether and phosphorus pentoxide
(SCHRAMM et al., 1962). Since this mixture of phosphate esters is
probably without any prebiologic significance and its reactions have
been discussed in detail in several of the references in the supplementary bibliography, nothing more will be said about it in this
article.

B. Replication

Virtually the only concerted effort directed toward the elucidation of
template-directed syntheses of oligonucleotides has been carried out
by ORGEL and his associates at The Salk Institute of Biological
Studies. Their investigations have been summarized well very recently
(ORGEL and LOHRMANN, 1974). Most of their work has resulted in un-
naturally linked dimers and trimers. The most successful results
under plausible prebiotic conditions involved the self-condensation of
adenosine cyclic 2',3'-phosphate on a polyuridylic acid template (RENZ
et al., 1971). The major products were adenosine 2'-and 3'-phosphates
produced by direct hydrolysis. A maximum of 20% of the dinucleotide
was produced along with a trace of trimer. The proportion of the un-
naturally 2',5'-linked product exceeded 97%.

IV. Terrestrial Life: Chance or Circumstance?

A. Chiral Homogeneity

Optical activity and terrestrial life seem to be conjointly united.
The property of optical activity is characteristic of biomolecules
that have a center of asymmetry, and it is due to the chirality or
handedness of the asymmetric molecule. Chirality is a structural
term, whereas optical activity is an operational term dependent on the
predominance of a unique chirality in a molecular population. Asymmetric biologic molecules are, with rare exception, all of the same
chirality and are, therefore, optically active. Just why asymmetric
biomolecules derived from living organisms should be of one homogeneous chirality and, hence, optically active, is a question that has
intrigued scientists for over a century. In 1860, Pasteur characterized "the molecular asymmetry of natural organic products as the great
characteristic which establishes perhaps the only well-marked line of
demarcation that can at present be drawn between the chemistry of
[nonbiologically derived] matter and the chemistry of living matter"
(JAPP, 1898). A thorough review of the problem has been presented by
BONNER (1972).

1. *The Statistical Viewpoint*

At the turn of the century, a spirited polemic on stereochemistry and vitalism (JAPP, 1898) evoked the first purely statistical explanation for the origin of optical activity (PEARSON, 1898a,b). RITCHIE (1947), however, objected that "the statistical explanation...has the weakness that, unless we assume that all living matter can be referred back to one single original microscopic particle with a dextro or levo asymmetry, this unilateral stereochemistry of living matter is difficult to explain." In other words, if optical activity did originate once in one microcosmic environment with molecules produced by a statistical fluctuation favoring one unique chirality, then these same statistical fluctuations could have given rise in another microenvironment to molecules of the opposite chirality.

2. *Asymmetric Physicochemical Forces*

PASTEUR (1848) was the first to seek experimental evidence for the origin of the unique chirality of biomolecules through the mediation of asymmetric geophysical forces. CURIE (1894), in commenting on Pasteur's unsuccessful experiments, pointed out that, among the few truly asymmetric forces in nature, circularly polarized light was the most common. VAN'T HOFF (1894) then suggested that optically active substances in nature might be formed "by transformations...which occur through the action of right- or left-circularly polarized light." Several degradations of chiral mixtures and several absolute asymmetric syntheses have been carried out by various investigators under the influence of circularly polarized light (BONNER, 1972). The major criticisms of these syntheses have been centered upon (1) the slight extent to which right-circularly polarized light occurs in nature, and (2) the purportedly trivial optical enrichment afforded by asymmetric photochemical processes.

The most ingenious physical explanation for the universal existence of optically active organic compounds in the biosphere that is currently under investigation involves the consequences of parity violation during the β-decay of radioactive isotopes (LEE and YANG, 1956). VESTER (1957) suggested an intrinsic causal relationship between the unique chirality associated with β-decay and the unique chirality prevailing in molecules associated with living matter. However, since interactions are negligibly small when energy levels are widely separated, ULBRICHT (1959) proposed that the longitudinally polarized β rays produced circularly polarized photons which then mediate the absolute asymmetric syntheses or degradations of chiral molecules. ULBRICHT and VESTER (1962) calculated that even under favorable circumstances such a mechanistic scheme might not lead to measurable optical activity in reasonable time spans.

The first positive experimental evidence that was offered for this mechanism was presented by GARAY (1968). Dilute, sterile, slightly basic solutions of tyrosine were exposed to sunlight while being bombarded with β-particles from strontium-90 that had been added to the solutions. After 18 months of observation based upon differences in ultraviolet absorption curves, GARAY reported that the D-isomer had been decomposing at a faster rate than the L-isomer. BONNER (1972) has criticized this report on the grounds that differences in the shape of absorption curves do not constitute a demonstration of induced optical activity. In addition, BONNER and FLORES (1974) have been unable to reproduce the GARAY experiments.

BONNER (1974) is involved at the present time in some long-range experiments in which D-, L-, and DL-amino acids, both as solids and as salts in aqueous solution, are being irradiated by a β-ray source of 61,700 Curies of strontium-90. The experiment will span several years, and although the results thus far show some fluctuations toward unique chirality, the data were not of sufficient statistical significance to obviate their being introduced by experimental error. Recently, BONNER et al. (1975) presented the first demonstration of the potential validity of the Vester-Ulbricht mechanism for the abiotic origin of optical activity. Employing a beam of 120-keV longitudinally polarized electrons from a linear accelerator, they were able to show that natural "left-handed" electrons produced an asymmetric degradation of D, L-leucine favoring the formation of an excess of the naturally occurring L-leucine enantiomer. FLORES (unpublished data, 1975) has just completed during the writing of this article a set of experiments which indicate that right circularly polarized light at 2128 Å preferentially photolyzes the D-enantiomer of D, L-leucine. The relationship between the latter and former experiments has not as yet been rationalized. Nonetheless, the question of whether natural asymmetric forces were of sufficient magnitude to account for the optical activity of biomolecules can be resolved only through a study of amplification mechanisms.

3. The Biotic Viewpoint

Biotic theories of the origin of optical activity assume that the very simplest life may have originated on a racemic basis and that optical activity gradually and subsequently emerged as life-forms developed. WALD (1975) has made a very detailed critique of abiotic origins. His main points are (1) restricted conditions, (2) poor yields, and (3) the tendency for only local and temporary asymmetry. WALD argued that adoption of a coiled α-helical secondary structure for proteins played an integral part in the formation and maintenance of primitive proteins, and that this structure in turn is most effectively promoted by the utilization of amino acids of a single configuration. Although this contention, that the α-helical structure is promoted by monomers of a single configuration, has been substantiated for high molecular weight polymers, STEINMAN (1967) has reported that there is little stereoselectivity at the oligopeptide level.

The difficulty with the biotic theories for the origin of optical activity is that one must assume that, since there was an equal opportunity for two populations of organisms to arise, each one having the mirror-image chirality of the other, then chance or some asymmetric geophysical force or chemical agent must finally have interceded in their development. The biotic viewpoint merely places the problem in a higher evolutionary hierarchy. Formally there are only two separate choices for the origin of optical activity: chance or asymmetric forces.

B. Evolutional Directiveness

1. *The Delimitation of Life*

Despite the existence of a theoretical biology, there appears to exist
as yet no satisfactory definition for life or even a delimitation of
life from nonlife (KEOSIAN, 1974). The reason for this is that any
definition or delimitation of life is an experiential axiom and cannot
be formulated as a logical conclusion. It has been suggested that the
terms "life" and "living" are meaningless (PIRIE, 1937) and that it
is useless to delimit life from nonlife since the transformation of
inanimate to animate matter probably proceeded via a series of hierarchical levels of organization (ROSEN, 1973: KEOSIAN, 1974). Nonetheless, definitions and delimitations for life appear now and then
in the scientific literature based either on information theory or
genetic replication. Their generalization and application has been
impeded because information, in its current theoretic usage, does not
define "value" (BRILLOUIN, 1962), and replication is but one aspect of
survival.

If it is allowed that Darwinian evolution is the most distinguishing
feature of the biosphere, then "value" can be defined in terms of
survival. Information impinges on a system in the form of mass or
energy, and "value" is the informational parameter that represents
survival benefit to the system for its existence as a structuralized
particle-matrix (GABEL, 1973). Structuralization, in this case, is
explicitly defined as organizational schemata that endue survival
potential. So structured, the system's "life" does not seem to be
dependent upon the arbitrary decision of an observer. This delimitation of life on the basis of the system's responsibility for the
primacy of its own survival, whether the system under consideration is
a species, population, or individual, is still an axiom and can be
subjected only to disputation; but it is, at least, an evolutional
delimitation that relates the system to impinging environmental information.

2. *Chemical Selectivity and the Genetic Code*

It is reasonable to assume that the genetic code had its origin in
prebiotic chemical selectivity. Whatever physicochemical phenomena
were responsible for this "recognition" of one chemical by another,
the selection process undoubtedly began at a simple level of organization. The rationale is quite fundamental. If the bulk of available
protobiotic matter had been polymerized to morphologically distinct
structures very early in the course of chemical evolution, then there
would have been numerically fewer chemical entities that could interact with each other. The probability of specific interactions that
could lead to any specific recognition apparatus would necessarily be
diminished. On the other hand, there would be a greater probability
for specific interaction at the monomeric (or oligomeric) level simply
because there would have been more monomers than polymers for any
given mass of protobiotic matter. Since biologic function and structure are inextricably intertwined, the origin of the genetic code and
its concomitant translation process could not have had its roots in

the interaction of preformed macromolecules of morphologic dimensions, for this would imply that structure arose independently of function. Within this context, the proteinoid microsphere theory of the origin of life (FOX et al., 1971a,b) is completely inadequate and is certainly not compatible with actualism.

3. *The Excitability Phenomenon*

One of the most overt but often ignored questions raised by the problem of the origin of life is: why did it all happen? ROSEN (1973) maintains that any dynamic model of evolution that does not incorporate a principle of function-change cannot explain the *raison d'etre* of evolution. PATTEE (1973) has long argued that structure and function cannot be understood independently of each other and still purport to explain the origin of life. WOESE (1972), in his chapter on the genetic code in *Exobiology*, underscores the fact that the genetic apparatus contains within itself no impetus for change and that it should be considered properly as a repository of information.

OPARIN (1974) pointed out recently that the non-Darwinian evolution of organized systems that preceded recognizable biologic forms was governed by a responsiveness to the environment by individual entities rather than by populations. Although OPARIN's coacervate model would be responsive to its aqueous environment, its ability to maintain its structural integrity while being impinged by environmental information would be dependent on the composition (material and spatial) of the coacervate drop. Even then, maintenance of structural integrity alone is not evolution. For evolution to occur, the structural model must in some way utilize the impinging information to increase its responsiveness and impart stability to whatever *principles* govern its structural *pattern*.

It is within the framework of actualism to examine trends in the contemporary biosphere in order to uncover these principles and this pattern. The most striking evolutionary trend of multiorgan systems, invertebrates as well as vertebrates, is the tendency toward cephalization. This tendency toward cephalization indicates some impetus toward increased development of excitability as evolution proceeds. The survival benefit of excitability to a biologic system, as a structuralized particle-matrix would disintegrate, because it is the information, in itself, that causes disruption.

During the course of a critical examination of observations of phenomena associated with excitable tissues (ABOOD, 1959; ABOOD et al., 1962, 1964), it was recognized that many of the properties of a viable neuronal membrane would also be present in a macromolecular polyphosphate coordination complex in which alkaline earth metals would have a cross-linking function and in which alkali metals would serve as accessory counterions (ABOOD and GABEL, 1965). Biologically excitable tissues (neural or muscular) are those in which diminutions in the cellular membrane potential are transmitted in a perpendicular direction with respect to the potential gradient of the membrane. These observations (Fig. 2) can be summarized as follows:

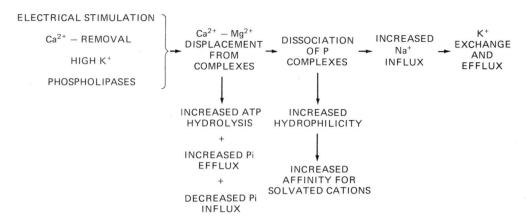

ELECTRICAL STIMULATION

Ca²⁺ − REMOVAL

HIGH K⁺

PHOSPHOLIPASES

Fig. 2. The events associated with depolarization of the excitable membrane (ABOOD and GABEL, 1965)

1. Depolarization (a diminution of membrane potential), whether produced by high external K⁺, displacement of Ca2+, or electrical
   stimulus, produces an increased efflux of Pi (orthophosphate) and
   ATP from excitable tissues.
2. This efflux is a function of the degree of depolarization.
3. Depolarization results in an inhibition of Pi influx and, secondarily, incorporation of Pi into ATP.
4. The impairment of Pi uptake closely parallels the decreasing membrane potential resulting from an increased concentration of external K⁺.
5. From a determination of thermodynamic constants for Pi efflux and
   Pi incorporation, it can be inferred that the two processes are
   somewhat related and that they are regulated both enzymatically
   and nonenzymatically.
6. The resting potential of a nonfiring membrane appears to be dependent upon the availability of ATP to the membrane.

The further development of concepts based upon these observations led
to a hypothesis which envisaged that linear polyphosphates leached
from primordial igneous rocks formed a macromolecular coordination
complex with the cations of sea water which, by virtue of its structure and chemical properties, would be excitable and would serve as
the evolutionary template for biopolymers and biologic membranes
(GABEL, 1965). The hypothesis conforms with actualism inasmuch as
excitation-relaxation phenomena characterize natural processes at all
hierarchical levels (RAINWATER et al., 1947; PRINGSHEIM, 1949; SMITH,
1970; GABEL, 1973), and the chemistry (SEMENOV et al., 1962; BELOV
and ORGANOVA, 1962; GRIFFITH et al., 1976) and biology (NOVELLI, 1967;
VAN STEVENINCK and BOOIJ, 1964; DEIERKAUF and BOOIJ, 1968; HAROLD,
1966; GABEL and THOMAS, 1971; KULAEV, 1971; GLONEK et al., 1971) that
are involved are in evidence today. The *principles* governing a structural pattern for this macromolecular coordination complex would be
the incorporation, transmittance, and utilization of environmental
information; the *pattern*, or in OPARIN's (1974) words "purposeful

activity," would be the entire spectrum of actions from responsivity to awareness; and the *phenomenon* would be excitability.

V. Conclusion

The investigation of chemical evolution is now in a state of flux. From the appearance of life on earth (SCHOPF, 1975) to the nature of the primitive atmosphere—many concepts are under reconsideration. Even the concept of the evolutionary transformation of inanimate matter to animate matter, as a result of the publication of the works of Teilhard DE CHARDIN, can now be found in the publications of religious bookstores (KOPP, 1964). MARX devitalized animate matter whereas Teilhard DE CHARDIN extended vitalism to inanimate matter. If the statement attributed to Harlow SHAPLEY: "We are kin to the stars and cousin to the stones," were to be taken out of context, it would be difficult to decide whether it had been inspired by Marxism or Teilhardism.

References

ABELSON, P.H.: Amino acids formed in primitive atmospheres. Science 124, 935 (1956).

ABELSON, P.H.: Chemical events on the primitive earth. Proc. Nat. Acad. Sci. 55, 1365-1372 (1966).

ABOOD, L.G.: Neuronal metabolism. In: Handbook of Physiology (Eds. J. FIELD, V. HALL) vol. 3, chapter 75, pp. 1815-1826. Washington, D.C.: Am. Physiol. Soc. 1959.

ABOOD, L.G., GABEL, N.W.: Relationship of calcium and phosphates to bioelectric phenomena in the excitatory membrane. Perspectives in Biology and Medicine 9, 1-12 (1965).

ABOOD, L.G., KOKETSU, K., MIYAMOTO, S.: Outflux of various phosphates during membrane depolarization of excitable tissues. Am. J. Physiol. 202, 469-474 (1962).

ABOOD, L.G., KOYAMA, I., THOMAS, V.: Relationship of depolarization to phosphorus metabolism and transport in excitable tissues. Am. J. Physiol. 207, 1435-1440 (1964).

ALBERS, R.W.: Biochemical aspects of active transport. Ann. Rev. Biochem. 36, 727-756 (1967).

ANBAR, M.: Cavitation during impact of liquid water on water: geochemical implications. Science 161, 1343-1344 (1968).

ANDERS, E., HAYATSU, R., STUDIER, M.H.: Organic compounds in meteorites. Science 182, 781-790 (1973).

ANDERS, E., HAYATSU, R., STUDIER, M.H.: Catalytic reactions in the solar nebula: implications for interstellar molecules and organic compounds in meteorites. Origins of Life 5, 57-67 (1974).

BARTH, C.: Seminar at Martin Marietta Corp., Denver, Colorado, June 15, 1970, quoted by T.A. PERLS.

BELOV, N.V., ORGANOVA, N.I.: Crystallochemistry and mineralogy of lomonosovite as related to its crystal structure. Geokhimiya, No. 1, 6-14 (1962); Chem. Abstr. 56, 15177a (1962).

BERNAL, J.D.: The Physical Basis of Life. London: Routledge and Paul 1951.

BITZ, M.C., NAGY, B.: Ozonolysis of polymer-type material in coal, kerogen and in the Orgueil meteorite. Proc. Nat. Acad. Sci. 56, 1383-1390 (1966).

BITZ, M.C., NAGY, B.: Analysis of bituminous coal by a combined method of ozonolysis, gas chromatography, and mass spectrometry. Anal. Chem. 39, 1310-1313 (1967).

BONNER, W.A.: Origins of molecular chirality. In: Exobiology (Ed. C. PONNAMPERUMA), pp. 170-234. Amsterdam: North-Holland 1972.

BONNER, W.A.: Experiments on the origin of molecular chirality by parity nonconservation during β-decay. J. Mol. Evol. 4, 23-39 (1974).

BONNER, W.A., FLORES, J.J.: Experiments on the origins of optical activity. In: Cosmochemical Evolution and the Origins of Life (Eds. J. ORO, S.L. MILLER, C. PONNAMPERUMA, R.S. YOUNG), pp. 187-194. Dordrecht (The Netherlands): D. Reidel Publ. Co. 1974.

BONNER, W.A., VAN DORT, M.A., YEARIAN, M.R.: The asymmetric degradation of D, L-leucine with longitudinally polarized electrons. Nature (Lond.), in press (1975).

BRANCH, G.E.K.: The free energy of formation of formic acid. J. Am. Chem. Soc. 37, 2316-2326 (1915).

BRILLOUIN, L.: Science and Information Theory, 2nd ed. New York: Academic Press 1962.

BROWN, H.: Rare gases and the formation of the earth's atmosphere. In: The Atmospheres of the Earth and Planets (Ed. G.P. KUIPER), 2nd ed., pp. 258-266. Chicago: University Press 1952.

BUHL, P.: An investigation of organic compounds in the Mighei meteorite. Ph.D. dissertation, Dept. of Chemistry, University of Maryland, College Park 1975.

BUKOWSKI, A., POREJKO, S.: The polymerization of carbon suboxide. Wiad. Chem. 23, 679-697 (1969).

CALVIN, M.: Chemical evolution and the origin of life. Am. Sci. 44, 248-263 (1956).

CALVIN, M.: Round trip from space. Evolution 13, 362-367 (1959).

CALVIN, M.: Chemical evolution. In: Condon Lectures, Oregon State Board of Higher Education. Eugene, Oregon: Univ. Oregon Press 1961.

CAMERON, A.G.W.: Formation of the outer planets. Space Sci. Revs. 14, 383-391 (1973a).

CAMERON, A.G.W.: Accumulation processes in the primitive solar nebula. Icarus 18, 407-415 (1973b).

CHUNG, N.M., LOHRMANN, R., ORGEL, L.E., RABINOWITZ, J.: The mechanism of the trimetaphosphate-induced peptide synthesis. Tetrahedron Letters 27, 1205-1210 (1971).

CRONIN, J.R., MOORE, C.B.: Amino acid analyses of the Murchison, Murray, and Allende carbonaceous chrondrites. Science 172, 1327-1329 (1971).

CURIE, P.: 1894, quoted by E.I. KLABUNOVSKII, Asymmetric synthesis, p. 197. Berlin: Veb. Deut. Verlag der Wissenschaften 1963.

DEIERKAUF, F.A., BOOIJ, H.L.: Changes in the phosphatide pattern of yeast cells in relation to active carbohydrate transport. Biochim. Biophys. Acta 150, 214-225 (1968).

FERRIS, J.P., CHEN, C.T.: Chemical evolution. XXVI. Photochemistry of methane, nitrogen, and water mixtures as a model for the atmosphere of the primitive earth. J. Am. Chem. Soc. 97, 2962-2967 (1975).

FOX, S.W.: The proteinoid theory of the origin of life and competing ideas. Am. Biol. Teach. 36, 161-181 (1974a).

FOX, S.W.: Origins of biological information and the genetic code. Mol. Cell. Biochem. (Enzymologia) 3, 129-132 (1974b).

FOX, S.W., HARADA, K.: Thermal copolymerization of amino acids to a product resembling protein. Science 128, 1214 (1958).

FOX, S.W., HARADA, K., KRAMPITZ, G., MUELLER, G.: Chemical origins of cells. Chem. Eng. News 48, No. 26, 80-94 (1970); *ibid.*, 49, No. 50, 46-53 (1971a).

FOX, S.W., YUKI, A., WAEHNELDT, T.V., LACEY, J.C., Jr.: The primordial sequence, ribosomes, and the genetic code. In: Molecular Evolution. I. Chemical Evolution and the Origin of Life (Eds. R. BUVET, C. PONNAMPERUMA), pp. 252-262. Amsterdam: North Holland 1971b.

GABEL, N.W.: Excitability and the origin of life: a hypothesis. Life Sci. 4, 2085-2096 (1965).

GABEL, N.W.: Could those rapidly exchangeable phosphoproteins be polyphosphate-protein complexes? Perspectives in Biology and Medicine 15, 640-643 (1972).

GABEL, N.W.: Abiogenic aspects of biological excitability. A general theory for evolution. In: Biogenesis-Evolution-Homeostasis (Ed. A. LOCKER), pp. 85-91. Heidelberg-Berlin: Springer 1973.

GABEL, N.W.: Polyphosphate as a possible precursor of primordial oligonucleotides. In preparation, 1976.

GABEL, N.W., PONNAMPERUMA, C.: A model for the primordial origin of monosaccharides. Nature (Lond.) 216, 453-457 (1967).

GABEL, N.W., THOMAS, V.: Occurrence of inorganic polyphosphates in excitable tissue. Fed. Proc. 28, 2663 (1969).

GABEL, N.W., THOMAS, V.: Evidence for the occurrence of inorganic polyphosphates in vertebrate tissue. J. Neurochem. 18, 1229-1242 (1971).

GARAY, A.S.: Origin and role of optical isomery in life. Nature (Lond.) 219, 338-340 (1968).

GARRISON, W.M., BENNETT, W., COLE, S.: Synthesis of products of higher molecular weight in the radiolysis of aqueous solutions of formic acid. Radiat. Res. 9, 647-659 (1958).

GARRISON, W.M., HAYMOND, H.R., MORRISON, D.C., WEEKS, B.M., GILE-MELCHERT, J.: High-energy helium-ion irradiation of aqueous acetic acid solutions. J. Am. Chem. Soc. 75, 2459-2464 (1953).

GARRISON, W.M., MORRISON, D.C., HAMILTON, J.G., BENSON, A., CALVIN, M.: Reduction of carbon dioxide in aqueous solutions by ionizing radiation. Science 114, 416-478 (1951).

GARRISON, W.M., MORRISON, D.C., HAYMOND, H.R., HAMILTON, J.G.: High-energy helium-ion irradiation of formic acid in aqueous solution. J. Am. Chem. Soc. 74, 4216-4223 (1952).

GETOFF, N.G., SCHOLES, G., WEISS, J.: Reduction of carbon dioxide in aqueous solutions under the influence of radiation. Tetrahedron Letters No. 18, 17-23 (1960).

GLONEK, T., KLEPS, R.A., GRIFFITH, E.J., MYERS, T.C.: Phosphorus-31 nuclear magnetic resonance studies on condensed phosphates. I. Some factors influencing the phosphate middle group chemical shift. Phosphorus, in press (1975).

GLONEK, T., LUNDE, M., MUDGETT, M., MYERS, T.C.: Biological polyphosphate studied through the use of phosphorus-31 nuclear magnetic resonance. Arch. Biochem. Biophys. 142, 508-513 (1971).

GRIFFITH, E.J.: Phosphate solubility in natural water systems. In: Proc. 14th Water Quality Conference, pp. 115-118. Urbana, Ill.: Univ. Ill. Press 1972.

GRIFFITH, E.J., PONNAMPERUMA, C., GABEL, N.W.: Phosphorus: a bridge to life on the primitive earth. Science, in press (1976).

HALDANE, J.B.S.: The origin of life. Rationalist Annual 148, 3-10 (1928).

HARADA, K., IWASAKI, T.: Synthesis of amino acids from aliphatic carboxylic acids by glow discharge electrolysis. Nature (Lond.) 250, 426 (1974).

HAROLD, F.M.: Inorganic polyphosphates in biology: structure, metabolism and function. Bacteriol. Revs. 30, 772-794 (1966).

HARTECK, P., GROTH, W., FALTINGS, K.: Photochemistry of CO. Z. Elektrochem. 44, 621-643 (1938).

HEINRICH, M.R., ROHLFING, D.L., BUGNA, E.: The effect of time of heating on the thermal polymerization of L-lysine. Arch. Biochem. Biophys. 130, 441-448 (1969).

HODGSON, G.W., BAKER, B.L.: Porphyrins in meteorites: metal complexes in Orgueil, Murray, Cold Bokkeveld, and Mokoia carbonaceous chondrites. Geochim. Cosmochim. Acta 33, 943-958 (1969).

HOLLAND, H.D.: Models for the evolution of the earth's atmosphere. In: Petrologic Studies: A Volume to Honor A.F. Buddington, pp. 447-477. Geol. Soc. Am. 1962.

HOROWITZ, N.H.: The evolution of biochemical syntheses. Proc. Nat. Acad. Sci. 31, 153-157 (1945).

HUBBARD, J.S., HARDY, J.P., HOROWITZ, N.H.: Photocatalytic production of organic compounds from CO and H_2O in a simulated Martian atmosphere. Proc. Nat. Acad. Sci. 68, 574-578 (1971).

HULSHOF, J., PONNAMPERUMA, C.: Prebiotic condensation reactions in an aqueous medium: a review of condensing agents. Origins of Life, in press (1975).

JAPP, F.R.: Stereochemistry and vitalism. Nature (Lond.) 58, 452-454 (1898).

KEOSIAN, J.: Life's beginnings origin or evolution. Origins of Life 5, 285-293 (1974).

KOPP, J.V.: Teilhard de Chardin: a new synthesis of evolution. Glen Rock, New Jersey: Paulist Press 1964.

KULAEV, I.S.: Inorganic polyphosphates in evolution of phosphorus metabolism. In: Molecular Evolution. I. Chemical Evolution and the Origin of Life (Eds. R. BUVET, C. PONNAMPERUMA), pp. 458-465. Amsterdam: North-Holland 1971.

KVENVOLDEN, K.A., LAWLESS, J., PERING, K., PETERSON, E., FLORES, J., PONNAMPERUMA, C., KAPLAN, I.R., MOORE, C.: Evidence for extraterrestrial amino acids and hydrocarbons in the Murchison meteorite. Nature (Lond.) 228, 923-926 (1970).

LANCET, M.S.: Carbon isotope fractionations in the Fischer-Tropsch reaction and noble gas solubilities in magnatite. Ph.D. dissertation, Univ. Chicago 1972.

LANCET, M.S., ANDERS, E.: Carbon isotope fractionation in the Fischer-Tropsch synthesis and in meteorites. Science 170, 980-982 (1970).

LAWLESS, J.G., FOLSOME, C.E., KVENVOLDEN, K.A.: Organic matter in meteorites. Sci. Am. 226, 38-46 (1972).

LAWLESS, J.G., KVENVOLDEN, K.A., PETERSON, E., PONNAMPERUMA, C.,
  MOORE, C.: Amino acids indigenous to the Murray meteorite. Science
  173, 626-627 (1971).
LEE, T.D., YANG, C.N.: The question of parity conservation in weak
  interactions. Phys. Rev. 104, 254-258 (1956).
LEWIS, B., VON ELBE, G.: Combustion, Flames, and Explosions of Gases,
  p. 753. New York: Academic Press 1951a.
LEWIS, B., VON ELBE, G.: op cit., p. 758, 1951b.
LIUTI, G., DONDES, S., HARTECK, P.: The photochemical separation of
  the carbon isotopes. In: Advances in Chemistry, Series No. 89,
  Isotope Effects in Chemical Processes, pp. 65-72. Washington, D.C.:
  Am. Chem. Soc. 1969.
LOHRMANN, R., ORGEL, L.E.: Prebiotic activation processes. Nature
  (Lond.) 244, 418-420 (1973).
MILLER, S.L.: Production of amino acids under possible primitive earth
  conditions. Science 117, 528-529 (1953).
MILLER, S.L.: Production of some organic compounds under possible
  primitive earth conditions. J. Am. Chem. Soc. 77, 2351-2361 (1955).
MILLER, S.L., ORGEL, L.E.: The Origins of Life on the Earth. Englewood
  Cliffs, New Jersey: Prentice-Hall 1973.
MILLER, S.L., UREY, H.C.: Organic compound synthesis on the primitive
  earth. Science 130, 245-251 (1959).
MILLS, J.A.: Association of polyhydroxy compounds with cations in
  solution. Biochem. Biophys. Res. Commun. 6, 418-421 (1961).
NEGRON-MENDOZA, A., PONNAMPERUMA, C.: In preparation (1976).
NOVELLI, G.D.: Amino acid activation for protein synthesis. Ann. Rev.
  Biochem. 36, 449-484 (1967).
OPARIN, A.I.: Proiskhozhdenie zhizni. Moscow: Moskovski Rabochii 1924.
OPARIN, A.I.: The Origin of Life, 2nd ed. New York: Dover 1938.
OPARIN, A.I.: The Chemical Origin of Life (trans. by A. SYNGE).
  Springfield, Illinois: Charles C. Thomas 1964.
OPARIN, A.I.: A hypothetic scheme for evolution of probionts. Origins
  of Life 5, 223-226 (1974).
ORGEL, L.E., LOHRMANN, R.: Prebiotic chemistry and nucleic acid
  replication. Accounts Chem. Res. 7, 368-377 (1974).
PASTEUR, L.: On the relations which can exist between crystalline
  form, chemical composition, and the sense of rotatory polarization.
  Ann. Chim. Phys. 24, 442-456 (1848).
PATTEE, H.: Experimental approaches to the origin of life problem.
  Adv. Enzymol. 27, 381-415 (1965).
PATTEE, H.H.: Physical problems of the origin of natural controls.
  In: Biogenesis-Evolution-Homeostasis (Ed. A. LOCKER), pp. 41-49.
  Heidelberg-Berlin: Springer 1973.
PEARSON, K.: Chance or vitalism. Nature (Lond.) 59, 495 (1898a).
PEARSON, K.: Asymmetry and vitalism. Nature (Lond.) 59, 30-31 (1898b).
PERING, K.L., PONNAMPERUMA, C.: Aromatic hydrocarbons in the Murchison
  meteorite. Science 173, 237-239 (1971).
PERLS, T.A.: Carbon suboxide on Mars: a working hypothesis. Icarus 14,
  252-264 (1971).
PIRIE, N.W.: The meaninglessness of the terms life and living. In:
  Perspectives in Biochemistry (Eds. J. NEEDHAM, D. GREEN), pp. 11-22.
  New York: Macmillan 1937.
PONNAMPERUMA, C., GABEL, N.W.: Current status of chemical studies on
  the origin of life. Space Life Sci. 1, 64-96 (1968).

PRINGSHEIM, P.: Fluorescence and Phosphorescence, p. 285. New York: Interscience Publishers 1949.

RABINOWITZ, J.: Peptide and amide bond formation in aqueous solutions of cyclic or linear polyphosphates as a possible prebiotic process. Helv. Chim. Acta 53, 1350-1355 (1970).

RABINOWITZ, J.: Sur le role du phosphore au cours de l'evolution chimique prebiologique. Chimia 26, 350-354 (1972).

RABINOWITZ, J., FLORES, J., KREBSBACH, R., ROGERS, G.: Peptide formation in the presence of linear or cyclic polyphosphates. Nature (Lond.) 224, 795 (1969).

RAFF, R.A., MEABURN, G.M.: Photochemical reaction mechanisms for production of organic compounds in a primitive earth atmosphere. Nature (Lond.) 221, 459-460 (1969).

RAINWATER, L.J., HAVENS, W.W., WU, C.S., DUNNING, J.R.: Slow neutron velocity spectrometer studies. I. Cd, Ag, Sb, Ir, Mn, Phys. Revs. 71, 65-79 (1947).

RENDLEMAN, J.A., Jr.: Complexes of alkali metals and alkaline-earth metals with carbohydrates. Adv. Carbohydrate Chem. 21, 209-269 (1966).

RENZ, M., LOHRMANN, R., ORGEL, L.E.: Catalysts for the polymerization of adenosine cyclic 2'-,3'-phosphate on a poly (U) template. Biochim. Biophys. Acta 240, 463-471 (1971).

RITCHIE, P.D.: Recent views on asymmetric synthesis and related processes. Adv. Enzymol. 7, 65-78 (1947).

ROSEN, R.: On the generation of metabolic novelties in evolution. In: Biogenesis-Evolution-Homeostasis (Ed. A. LOCKER), pp. 113-123. Berlin-Heidelberg-New York: Springer 1973.

RUBEY, W.W.: Development of the hydrosphere and atmosphere with special reference to probable composition of the early atmosphere. In: Crust of the Earth, Spec. Paper 62, pp. 631-650. Washington D.C.: Geol. Soc. Am. 1955.

RUSSELL, H.N.: The Solar System and Its Origin. New York: Macmillan 1935.

RUTTEN, M.G.: The Geological Aspects of the Origin of Life on Earth. Amsterdam: Elsevier 1962.

RUTTEN, M.G.: The Origin of Life by Natural Causes. Amsterdam: Elsevier 1971.

SCHOPF, J.W.: Precambrian palebiology: problems and perspectives. Ann. Rev. Earth Planetary Sci. 3, 213-249 (1975).

SCHRAMM, G., GROTSCH, H., POLLMANN, W.: Non-enzymic synthesis of polysaccharides, nucleosides, and nucleic acids and the origin of systems capable of self-reproduction. Angew. Chem. 74, 53-59 (1962).

SEMENOV, E.I., ORGANOVA, N.I., KUKHARCHIK, M.V.: New data on minerals of the lomonosovite-murmanite group. Kristallografiya 6, 925-932 (1961); Chem. Abstr. 56, 11023d (1962).

SHKLOVSKII, I.S., SAGAN, C.: Intelligent Life in the Universe. San Francisco: Holden-Day 1966.

SIMONAITIS, R., HEICKLEN, J.: Mercury sensitized photolysis of nitrogen gas and carbon monoxide. J. Photochem. 1, 181-196 (1973).

SMITH, W.V.: Research management. Science 167, 957-959 (1970).

STEINMAN, G.: Stereoselectivity in peptide synthesis under simple conditions. Experientia 23, 177-178 (1967).

SWALLOW, A.J.: Radiation Chemistry of Organic Compounds, p. 242. New York: Pergamon Press 1960.

TANAKA, F.S., WANG, C.H.: Radiolysis of succinic acid in aqueous solution. Int. J. Appl. Radiat. Isot. 18, 761-772 (1967).

ULBRICHT, T.L.V.: Asymmetry: the non-conservation of parity and optical activity. Quart. Rev. 13, 48-60 (1959).

ULBRICHT, T.L.V., VESTER, F.: Attempts to induce optical activity with polarized β-radiation. Tetrahedron Letters 18, 629-637 (1962).

UREY, H.C.: The Planets: Their Origin and Development. London: Oxford Univ. Press 1952.

VAN STEVENINCK, J. BOOIJ, H.L.: The role of polyphosphates in the transport mechanism of glucose in yeast cells. J. Gen. Physiol. 48, 43060 (1964).

VAN'T HOFF, J.H.: The Arrangement of Atoms in Space, 2nd ed., p. 30. 1894; 3rd ed., p. 8. London: Longmans, Green, and Co. 1908.

VAN TRUMP, J.E., MILLER, S.L.: Carbon monoxide on the primitive earth. Earth Planet. Sci. Lett. 20, 145-150 (1973).

VESTER, F.: Seminar at Yale University, February 7, 1957.

VINOGRADOV, A.P., TUGARINOV, A.I.: Geochronology of the Precambrian. Geokhimiya No. 723-731 (1961).

WALD, G.: The origin of optical activity. Ann. N.Y. Acad. Sci. 69, 352-357 (1957).

WALKER, J.C.G.: Implications for atmospheric evolution of the in-homogeneous accretion model of the origin of the earth. In: The Early History of the Earth (Ed. B.F. WINDLEY). New York: Wiley 1976.

WHELAND, G.W.: Advanced Organic Chemistry, 2nd ed., Chapter 1. New York: John Wiley and Sons 1949.

WIEGAND, W.J., NIGHAN, W.L.: Plasma chemistry of carbon dioxide-molecular nitrogen-helium discharges. Appl. Phys. Lett. 22, 583-586 (1973).

WOESE, C.R.: The emergence of genetic organization. In: Exobiology (Ed. C. PONNAMPERUMA), Chapter 9, pp. 301-341. Amsterdam: North-Holland 1972.

WRIGHT, A.N., WINKLER, C.A.: Active Nitrogen. New York: Academic Press 1968.

YOUNG, R.A., MORROW, W.: Formation of CN from CO and its excitation in active nitrogen. J. Chem. Phys. 60, 1005-1008 (1974).

Supplementary Bibliography

CAIRNS-SMITH, A.G.: The Life Puzzle. Edinburgh: Oliver and Boyd 1971.

CALVIN, M.: Chemical Evolution. New York: Oxford University Press 1969.

FOX, S.W., DOSE, K.: Molecular Evolution and the Origin of Life. San Francisco: Freeman 1972.

KENYON, D.H., STEINMAN, G.: Biochemical Predestination. New York: McGraw-Hill 1969.

KIMBALL, A.P., ORO, J., Eds.: Prebiotic and Biochemical Evolution. Amsterdam: North-Holland 1971.

MARGULIS, L., Ed.: Origins of Life. I. New York: Gordon and Breach 1970.

MARGULIS, L., Ed.: Origins of Life. II. New York: Gordon and Breach 1971.

ORGEL, L.E.: The Origins of Life. New York: Wiley 1973.

PONNAMPERUMA, C., Ed.: Exobiology. Amsterdam: North-Holland 1972.

PONNAMPERUMA, C.: The Origins of Life. London: Thames and Hudson 1972.

SCHWARTZ, A.W., Ed.: Theory and Experiment in Exobiology. Groningen (The Netherlands): Wolters-Noordhoff 1972.

ROHLFING, D.L., OPARIN, A.I., Eds.: Molecular Evolution Prebiological and Biological. New York: Plenum Press 1972.

WEST, M.W., PONNAMPERUMA, C.: Chemical evolution and the origin of life: a comprehensive bibliography. Space Life Sci. 2, 225-295 (1970).

WEST, M.W., GILL, E.D., PONNAMPERUMA, C.: Chemical evolution and the origin of life: bibliography supplement 1970. Space Life Sci. 3, 293-304 (1972).

WEST, M.W., GILL, E.D., SHERWOOD, B., KVENVOLDEN, K.A.: Chemical evolution and the origin of life: bibliography supplement 1972. Origins of Life 5, 507-527 (1974).

Subject Index

174

Progress in Molecular and Subcellular Biology

Editorial Board:
F.E. Hahn, H. Kersten,
W. Kersten, T.T. Puck,
G.F. Springer,
W. Szybalski, K. Wallenfels
Managing Editor:
F.E. Hahn

Volume 1
By B.W. Agranoff, J. Davies,
F.E. Hahn, H.G. Mandel,
N.S. Scott, R.M. Smillie,
C.R. Woese
32 figures. VII, 237 pages.
1969

Volume 1 contains contributions by Woese, Davies and Mandel who are concerned either theoretically or experimentally with the nature of the genetic code and its correct translation. Smillie and Scott write about the biosynthesis of chloroplasts and the photo-regulation of the underlying basic processes. Agranoff reviews critically current ideas concerning the molecular basis of higher nervous activities. The managing editor, Hahn, makes a critical analysis of the origin, conceptual content and probable future developments in molecular biology.

Volume 2
Proceedings of the Research Symposium on Complexes of Biologically Active Substances with Nucleic Acids and Their Modes of Action

Held at the Walter Reed Army Institute of Research, Washington, 16-19 March, 1970. 158 figures.
IX, 400 pages. 1971

These 28 contributions are a definitive account of the most recent research results on selected aspects of the title topic. The major topical subdivisions are: Antibiotics and Nucleic Acids — Antimalarials and Nucleic Acids — Alkaloids, Natural Polyamines and DNA — Intercalation into Supercoiled DNA — Synthetic Drugs and Dyes Binding to Nucleic Acids — Antimutagens — Carcinogens and Nucleic Acids — The Natural State of DNA. Among the contributers are D.M. Crothers, G.F. Gause, W. and H. Kersten, L.S. Lerman, H.R. Mahler, P.O.P. Ts'o, J. Vinograd, and M. Waring.

Volume 3
By A.S. Braverman,
D.J. Brenner, B.P. Doctor,
A.B. Edmundson, K. R. Ely,
M.J. Fournier, F.E. Hahn,
A. Kaji, C.A. Paoletti,
G. Riou, M. Schiffer,
M.K. Wood 58 figures. VII,
251 pages. 1973

The volume is concerned with the transcription and translation of genetic information and with products of both processes in prokaryotic and eukaryotic organisms. Reverse transcription is reviewed with respect to its theoretical importance to molecular biology and its practical importance to cancer research. Translation, i.e., protein biosynthesis, is analyzed in the light of

knowledge of numerous inhibitors of individual reaction steps in the overall sequence. The transcription of transfer RNA is reviewed in detail as one aspect of transcription of genetic information. Thalassemia and Bence-Jones proteins have been treated as selected examples of molecular pathology in eukaryotes. The nature of mitochondrial DNA in neoplastic cells is reviewed as one contribution to the molecular biology of cancer. The book signifies the advancement of molecular biology from the study of bacteria and their viruses to the consideration of events and entities in eukaryotic organisms.

Volume 4
By S. Bram, H.J. Witmer,
W. Lotz, M. Sussman,
D. Oesterhelt, P. Chandra,
L.K. Steel, U. Ebener,
M. Woltersdorf, H. Laube,
G. Will, R.A. Olsson,
R.E. Patterson
88 figures. 27 tables. XI,
251 pages. 1976

This volume contains seven articles which range from biophysical studies on the conformation of DNA, through papers on gene expression in bacterial viruses and cellular slime molds, to inhibitors of nucleic acid syntheses as antitumor agents and the role of adenosine as a regulator of coronary blood flow. It continues to combine carefully selected articles in the field of biomedicine with those in which life scientists are particularly interested.

Springer-Verlag
Berlin Heidelberg New York

**Springer-Verlag
Berlin
Heidelberg
New York**